Future Biotechnology Research on the International Space Station

Task Group for the Evaluation of NASA's Biotechnology Facility
for the International Space Station

Space Studies Board

Commission on Physical Sciences, Mathematics, and Applications

National Research Council

NATIONAL ACADEMY PRESS
Washington, D.C.

NOTICE: The project that is the subject of this report was approved by the Governing Board of the National Research Council, whose members are drawn from the councils of the National Academy of Sciences, the National Academy of Engineering, and the Institute of Medicine. The members of the task group responsible for the report were chosen for their special competences and with regard for appropriate balance.

Support for this project was provided by Contract NASW 96013 between the National Academy of Sciences and the National Aeronautics and Space Administration. Any opinions, findings, conclusions, or recommendations expressed in this material are those of the authors and do not necessarily reflect the views of the sponsor.

Images on the front cover:

Top left: Red blood cells in a cross section of a small blood vessel in the mouse lung. Scanning electron micrograph courtesy of Jacob Bastacky, Lawrence Berkeley National Laboratory.

Top right: A ligand complex, a designed, sequence-specific minor groove binding compound (Geierstanger et al., 1996). Image courtesy of David Wemmer, University of California at Berkeley.

Bottom left: Nitrogen Regulatory Protein C (NtrC), a bacterial transcription factor regulated by phosphorylation (Volkman et al., 1995). Image courtesy of David Wemmer, University of California at Berkeley.

Bottom right: The apicoplast is a nonphotosynthetic plastid, acquired by endosymbiosis of a eukaryotic alga and retention of the algal chloroplast. This essential organelle is found in all members of the phylum Apicomplexa, including malaria parasites and the AIDS pathogen *Toxoplasma gondii* (shown), providing a promising target for therapeutic drug design (Köhler et al., 1997). Electron micrograph courtesy of David Roos and Lewis Tilney, University of Pennsylvania.

International Standard Book Number 0-309-06975-0

Copies of this report are available free of charge from:

Space Studies Board
National Research Council
2101 Constitution Avenue, NW
Washington, DC 20418

Copyright 2000 by the National Academy of Sciences. All rights reserved.

Printed in the United States of America

THE NATIONAL ACADEMIES

National Academy of Sciences
National Academy of Engineering
Institute of Medicine
National Research Council

The **National Academy of Sciences** is a private, nonprofit, self-perpetuating society of distinguished scholars engaged in scientific and engineering research, dedicated to the furtherance of science and technology and to their use for the general welfare. Upon the authority of the charter granted to it by the Congress in 1863, the Academy has a mandate that requires it to advise the federal government on scientific and technical matters. Dr. Bruce M. Alberts is president of the National Academy of Sciences.

The **National Academy of Engineering** was established in 1964, under the charter of the National Academy of Sciences, as a parallel organization of outstanding engineers. It is autonomous in its administration and in the selection of its members, sharing with the National Academy of Sciences the responsibility for advising the federal government. The National Academy of Engineering also sponsors engineering programs aimed at meeting national needs, encourages education and research, and recognizes the superior achievements of engineers. Dr. William A. Wulf is president of the National Academy of Engineering.

The **Institute of Medicine** was established in 1970 by the National Academy of Sciences to secure the services of eminent members of appropriate professions in the examination of policy matters pertaining to the health of the public. The Institute acts under the responsibility given to the National Academy of Sciences by its congressional charter to be an adviser to the federal government and, upon its own initiative, to identify issues of medical care, research, and education. Dr. Kenneth I. Shine is president of the Institute of Medicine.

The **National Research Council** was organized by the National Academy of Sciences in 1916 to associate the broad community of science and technology with the Academy's purposes of furthering knowledge and advising the federal government. Functioning in accordance with general policies determined by the Academy, the Council has become the principal operating agency of both the National Academy of Sciences and the National Academy of Engineering in providing services to the government, the public, and the scientific and engineering communities. The Council is administered jointly by both Academies and the Institute of Medicine. Dr. Bruce M. Alberts and Dr. William A. Wulf are chairman and vice chairman, respectively, of the National Research Council.

TASK GROUP FOR THE EVALUATION OF NASA'S BIOTECHNOLOGY FACILITY FOR THE INTERNATIONAL SPACE STATION

PAUL B. SIGLER, Yale University, *Chair**
ADELE L. BOSKEY, Hospital for Special Surgery, New York
NOEL D. JONES, Molecular Structure Corporation and Eli Lilly (retired)
JOHN KURIYAN, Rockefeller University
WILLIAM M. MILLER, Northwestern University
MICHAEL L. SHULER, Cornell University
GARY S. STEIN, University of Massachusetts Medical School**
BI-CHENG WANG, University of Georgia

ELIZABETH L. GROSSMAN, Program Officer
CARMELA J. CHAMBERLAIN, Senior Project Assistant

---

*Deceased, January 11, 2000.
**Acting chair as of January 12, 2000.

SPACE STUDIES BOARD

CLAUDE R. CANIZARES, Massachusetts Institute of Technology, *Chair*
MARK R. ABBOTT, Oregon State University
FRAN BAGENAL, University of Colorado
DANIEL N. BAKER, University of Colorado
ROBERT E. CLELAND, University of Washington
MARILYN L. FOGEL, Carnegie Institution of Washington
BILL GREEN, former member, U.S. House of Representatives
JOHN H. HOPPS, JR., Morehouse College
CHRISTIAN J. JOHANNSEN, Purdue University
RICHARD G. KRON, University of Chicago
JONATHAN I. LUNINE, University of Arizona
ROBERTA BALSTAD MILLER, Columbia University
GARY J. OLSEN, University of Illinois, Urbana-Champaign
MARY JANE OSBORN, University of Connecticut Health Center
GEORGE A. PAULIKAS, The Aerospace Corporation
JOYCE E. PENNER, University of Michigan
THOMAS A. PRINCE, California Institute of Technology
PEDRO L. RUSTAN, JR., U.S. Air Force (retired)
GEORGE L. SISCOE, Boston University
EUGENE B. SKOLNIKOFF, Massachusetts Institute of Technology
MITCHELL SOGIN, Marine Biological Laboratory
NORMAN E. THAGARD, Florida State University
ALAN M. TITLE, Lockheed Martin Advanced Technology Center
RAYMOND VISKANTA, Purdue University
PETER W. VOORHEES, Northwestern University
JOHN A. WOOD, Harvard-Smithsonian Center for Astrophysics

JOSEPH K. ALEXANDER, Director

COMMISSION ON PHYSICAL SCIENCES, MATHEMATICS, AND APPLICATIONS

PETER M. BANKS, Veridian ERIM International, Inc., *Co-chair*
W. CARL LINEBERGER, University of Colorado, *Co-chair*
WILLIAM F. BALLHAUS, JR., Lockheed Martin Corp.
SHIRLEY CHIANG, University of California
MARSHALL H. COHEN, California Institute of Technology
RONALD G. DOUGLAS, Texas A&M University
SAMUEL H. FULLER, Analog Devices, Inc.
JERRY P. GOLLUB, Haverford College
MICHAEL F. GOODCHILD, University of California at Santa Barbara
MARTHA P. HAYNES, Cornell University
WESLEY T. HUNTRESS, JR., Carnegie Institution
CAROL M. JANTZEN, Westinghouse Savannah River Company
PAUL G. KAMINSKI, Technovation, Inc.
KENNETH H. KELLER, University of Minnesota
JOHN R. KREICK, Sanders, a Lockheed Martin Company (retired)
MARSHA I. LESTER, University of Pennsylvania
DUSA M. McDUFF, State University of New York at Stony Brook
JANET L. NORWOOD, Former Commissioner, U.S. Bureau of Labor Statistics
M. ELISABETH PATÉ-CORNELL, Stanford University
NICHOLAS P. SAMIOS, Brookhaven National Laboratory
ROBERT J. SPINRAD, Xerox PARC (retired)

MYRON F. UMAN, Acting Executive Director

Foreword

Under current NASA plans, investigations in the area of biotechnology will be a significant component of the life sciences research to be conducted on the International Space Station (ISS). They encompass work on cell science and studies of the use of microgravity to grow high-quality protein crystals. Both these subdisciplines are advancing rapidly in terrestrial laboratories, fueled by federal and industrial research budgets that dwarf those of NASA's life science program. Forging strong and fruitful connections between the space investigations and laboratory-bench biologists, a continual challenge for NASA's life sciences program, is thus of great importance to ensuring the excellence of ISS research.

This report evaluates the plan for NASA's biotechnology facility on the ISS and the scientific context that surrounds it, and makes recommendations on how the facility can be made more effective. In addition to questions about optimizing the instrumentation, the report addresses strategies for enhancing the scientific impact and improving the outreach to mainstream terrestrial biology. No major redirection of effort is called for, but collectively the specific, targeted changes recommended by the task group would have a major effect on the conduct of biotechnology research in space.

The sudden death of the task group's chair, Dr. Paul Sigler of Yale University, came when this document was nearly completed. Dr. Sigler was a world-renowned scientist; indeed he was a superb example of the terrestrial biologists who need to be drawn toward NASA's research program in larger numbers. His wisdom and insight are evident throughout the report.

Claude R. Canizares, *Chair*
Space Studies Board

Preface

This report was prepared by the Task Group for the Evaluation of NASA's Biotechnology Facility for the International Space Station (ISS) in response to a request from the National Aeronautics and Space Administration (NASA). The task group was formed to examine NASA's ongoing program on biotechnology in the microgravity environment, including the plans for relevant instrumentation to be installed on the ISS (a complete statement of task is provided in Appendix C). Previous National Research Council (NRC) reports have touched on issues related to NASA's biotechnology program and on equipment for the ISS. *Microgravity Research Opportunities for the 1990s* (NRC, 1995) offered recommendations about promising areas of potential research for NASA's microgravity sciences program; suggestions specifically for biotechnology work were included. *Future Materials Science Research on the International Space Station* (NRC, 1997) looked in detail at the Space Station Furnace Facility, which was designed to support and enable materials science research on the ISS, and at NASA's priorities and project selection process for materials science research. *A Strategy for Research in Space Biology and Medicine in the New Century* (NRC, 1998) provides a broad survey of opportunities and priorities for NASA research in a variety of areas, including cell biology. *Institutional Arrangements for Space Station Research* (NRC, 1999) discusses the formation of a nongovernmental organization designed to facilitate research on the ISS. This report, *Future Biotechnology Research on the International Space Station,* covers both scientific and technical questions related to NASA's future plans for biotechnology research on the ISS and has a scope similar to that of the materials science report.

NASA's biotechnology program consists of two distinct research areas: protein crystal growth and cell science. In the first area, NASA scientists and NASA-sponsored researchers study macromolecular crystallization processes and production of high-quality crystals in the microgravity environment. In the second area, research focuses on the study of cell and tissue culture, growth, and differentiation. The composition of the task group, which included researchers with expertise in macromolecular crystallography for structure determination, instrumentation for crystallization and X-ray diffraction, tissue engineering, and in vitro studies of cell differentiation and proliferation, reflected the dual nature of the NASA program. The task group also included a mix of scientists familiar with the constraints imposed by research in a microgravity environment and those with no previous experience in space-based experiments but with a broad-based understanding of the research communities' current issues and potential future needs. Brief biographical information about the task group members is available in Appendix B.

The task group held two meetings and two site visits over the course of the study to gather information on past results of NASA's program and on future plans for utilization of the ISS for biotechnology research. The task group has provided background on the NASA biotechnology program and a discussion of the scientific goals in each field in Chapter 1. In Chapter 2, the instrumentation planned for the ISS is described, and the task group comments on the equipment, on the hardware development process, and on effective use of the ISS. Finally, in Chapter 3, the task group examines NASA's approach to project selection and community outreach and offers suggestions for strengthening these processes.

This study was conducted under the auspices of the NRC's Space Studies Board and its staff, and the task group acknowledges this support. The task group also would like to thank the many people from NASA and other institutions who supplied extensive background materials, hosted site visits, and provided thorough presentations. Particular recognition is due to Steve Davison at NASA headquarters, Neal Pellis at the Johnson Space Center, and Craig Kundrot and Ron Porter at the Marshall Space Flight Center.

Acknowledgment of Reviewers

This report has been reviewed by individuals chosen for their diverse perspectives and technical expertise, in accordance with procedures approved by the National Research Council's (NRC's) Report Review Committee. The purpose of this independent review is to provide candid and critical comments that will assist the authors and the NRC in making the published report as sound as possible and to ensure that the report meets institutional standards for objectivity, evidence, and responsiveness to the study charge. The contents of the review comments and draft manuscript remain confidential to protect the integrity of the deliberative process. We wish to thank the following individuals for their participation in the review of this report:

Michael J. Bentenbaugh, Johns Hopkins University,
Kenneth I. Berns, University of Florida,
Shu Chien, University of California, San Diego,
Howard M. Einspahr, Bristol-Myers Squibb Pharmaceutical Research Institute,
Donald E. Ingber, Children's Hospital and Harvard Medical School,
Kenneth H. Keller, University of Minnesota, and
J. Keith Moffat, University of Chicago.

Although the individuals listed above have provided many constructive comments and suggestions, responsibility for the final content of this report rests solely with the authoring task group and the NRC.

Contents

EXECUTIVE SUMMARY 1

1 BACKGROUND AND SCIENTIFIC SCOPE OF NASA PROGRAMS 10
   Introduction, 10
   Protein Crystal Growth, 11
      The Significance of Crystallographic Resolution Limits, 11
      Goals and History of the NASA Protein Crystal Growth Effort, 12
      Results to Date: Examples of Successful Experiments and the
         Importance of Defining Controls, 13
      Potential Areas of Future Impact, 15
      Potential Benefits of the Space Station Platform, 17
      Potential for Interest from Commercial Entities, 18
   Cell Science, 18
      Goals and Potential Impacts of the NASA Cell Science Effort, 18
      Experimental Design and Instrumentation, 21
      Requirements for Interprogrammatic Coordination Within NASA, 22

2 INSTRUMENTATION 23
   Logistics for Using the International Space Station as a Biotechnology Research Platform, 23
   Protein Crystal Growth, 24
      The Hardware Development Process, 24
      Key Characteristics of Protein Crystal Growth Hardware on the ISS, 25
      The X-ray Crystallography Facility, 26
   Cell Science, 27
      Cell and Tissue Culture Hardware, 28
      Experiment Management, 32
      Storage, Transport, and Throughput of Samples, 35
   Overall Volume Allotment for Biotechnology Research on the ISS, 36

3	SELECTION AND OUTREACH	38

    Selection Process, Outreach Efforts, and Communication Among Program Participants, 38
        Improving the Dissemination of NRAs and NASA Program Results, 38
        Improving the Selection Process, 40
        Improving Connections to Relevant Communities and Attracting the Best Science, 41
        Coordination: Investigators and Operations Personnel, 42
    Protein Crystal Growth, 43
        The Guest Investigator Program, 43
        Funding Research on Biologically Challenging Problems, 44
    Cell Science, 46
        Cooperation with NASA's Life Sciences Division and with Other Federal Agencies, 46
        Resource Management and Communication in Times of Crisis, 46

BIBLIOGRAPHY	48

APPENDIXES

A	Hardware Available or in Development and Schedule for Biotechnology Research on the International Space Station	53
	Hardware for Protein Crystal Growth in Space, 53	
	Hardware for Cell Science in Space, 58	
	Schedule, 61	
B	Biographical Sketches of Task Group Members	62
C	Statement of Task	64
D	Glossary	65
E	Acronyms and Abbreviations	68

This report is dedicated to the memory of

Professor Paul B. Sigler
(1934 – 2000)

*a distinguished scholar and researcher whose leadership and
insight made him a vital contributor to this report.*

Executive Summary

BACKGROUND AND SCIENTIFIC SCOPE OF NASA PROGRAMS

The National Aeronautics and Space Administration (NASA) manages research programs in two areas of the rapidly expanding field of biotechnology: protein crystal growth and cell science. The protein crystal growth work focuses on using microgravity to produce higher quality macromolecular crystals for structure determination and on improving understanding of the crystal growth process. The cell science work focuses on basic research that contributes to understanding how the microgravity environment affects the fundamental behavior of cells, particularly in relation to tissue formation and the effects of space exploration on living organisms. The National Research Council's Task Group for the Evaluation of NASA's Biotechnology Facility for the International Space Station was formed to examine and evaluate the use of the International Space Station (ISS) as a platform for research in these two areas. In this report, the task group offers a variety of recommendations and suggestions for improving the NASA biotechnology research program. It believes these changes are necessary if the NASA program is to fulfill the potential for scientific discovery and impact that is also outlined in this report.

Protein Crystal Growth

The task group heard a great deal about experiments to date in NASA's macromolecular crystallography program. The results so far are inconclusive, and the impact of microgravity crystallization on structural biology as a whole has been extremely limited. At this time, one cannot point to a single case where a space-based crystallization effort was the crucial step in achieving a landmark scientific result. In many of the cases that have so far been listed as successful, the improvements obtained have been incremental rather than fundamental. In addition, the difficulty of mounting simultaneous efforts to produce the best possible crystals both on the ground and in space has limited the ability of researchers to make the comparisons between microgravity and Earth crystals that would be necessary to demonstrate that the microgravity environment can produce superior crystals.

Finding: The results from the collection of experiments performed on microgravity's effect on protein crystal growth are inconclusive. The improvements in crystal quality that have been observed are often only incremental, and the difficulty of producing the appropriate controls limits investigators' ability to definitively assess if improvements can be reliably credited to the microgravity environment. To date, the impact of microgravity crystallization on structural biology as a whole has been extremely limited.

Despite the lack of impact of microgravity research on structural biology up to now, there is reason to believe that the potential exists for crystallization in the microgravity environment to contribute to future advances in structure determination. Today's ground-based protein crystallization projects are increasingly sophisticated, and yet the diffraction characteristics of crystals of many important targets are still suboptimal. Improvements in diffraction that move a system from the margins of structure determination to well beyond that boundary will have a significant impact on the ability of the resulting structure to provide important insights into biological mechanisms. All research on protein crystallization in space has, up to now, been done under suboptimal conditions (short-duration experiments, insufficient vibration control, etc.), so the improved conditions for research provided by the ISS have the potential to produce much better results.

Finding: While enormous strides have been made in protein crystallization in the last decade, it is still the case that there are very important classes of compelling biological problems where the difficulty of obtaining crystals that diffract to high resolution remains the chief barrier to structural analysis of the crystals. It is here that the NASA program must look to maximize its impact.

In order to engage the research community, NASA must focus its support on programs that are developing technologically innovative equipment and engaging in the structure determination of crystals with important biological implications. While past NASA-supported research on the crystallization process has not been without value, NASA's priority should now be to resolve the community's questions about the usefulness of protein crystal growth in the microgravity environment for tackling important biological questions. Until the uncertainty about the value of space-based crystallization is resolved, a program of this fiscal magnitude is bound to engender resentment in the scientific community.

Although many pharmaceutical and biotechnology companies have participated in microgravity crystallization research, not one has yet committed substantial financial resources to the program. This is likely to remain the case until the benefits of microgravity can be convincingly documented by basic researchers and until facilities in space can handle greatly increased numbers of samples in a much more user friendly manner.

Cell Science

NASA's cell science program focuses on studying the influence of low gravity on fundamental cell biology as it relates to tissue formation, and on providing insight into the effects of microgravity on cell, tissue, and organ system function, especially as it might affect participants in space exploration.

Finding: It is appropriate for NASA to support a cell science program aimed at exploring the fundamental effects of the microgravity environment on biological systems at the cellular level. Results from such basic research experiments could have a significant impact on the fields of cell science and tissue engineering. However, the specific important questions within cell biology that can best be tackled on the ISS do not seem to have been defined yet. Narrowing the broad sweep of the current program may focus instrument development efforts and accelerate progress toward complete understanding of the effects of microgravity on specific biological phenomena.

A key to determining the success of cell science experiments in space will be designing appropriate controls for experiments. In space, cell cultures experience a low gravitational environment that reduces convection, buoyancy-driven flows, and sedimentation, and it is difficult to separate the various factors causing differences between space- and Earth-grown samples. In addition, the tremendous progress that has been made in three-dimensional tissue development on Earth, under unit gravity, provides a wide range of options for ground-based experiments that may produce results similar to those achieved in microgravity. To evaluate the relative merits of various experimental control groups, and also to enable the detailed evaluation of samples returned from space, it is important that quantitative measures of cell and tissue structure and function be developed and studied.

Finding: Appropriate experimental controls for space-based cell science experiments have not yet been determined. The best controls would be those that enable researchers to separate and investigate the multiple factors—including launch and reentry, effects of microgravity on the culture medium, and direct effects of microgravity on cellular behavior—that produce the changes observed in cells and tissues grown in space. Analytical techniques that measure the molecular mechanisms underlying cellular functions will be essential to provide data for comparing proposed experimental controls and quantifying the observed changes in cell and tissue samples.

At NASA, the work viewed by the task group was being carried out in the biotechnology section of the Microgravity Research Division. The themes of the cell science research under way in this program overlap with the scope of work ongoing in the NASA Life Sciences Division. The complementary nature of these two programs needs to be recognized so that NASA personnel and external researchers can take full advantage of the potential synergies. While there is already a sharing of flight hardware, a mechanism to establish projects that are jointly funded by the Life Sciences Division and the Microgravity Research Division should be considered.

Recommendation: The research strategies and projects of the cell science work in the biotechnology section of the Microgravity Research Division should be more closely coordinated with the work of NASA's Life Sciences Division to take advantage of overlapping work on bone and muscle constructs and of potential synergies between in vitro and in vivo research projects.

INSTRUMENTATION

The International Space Station (ISS) is currently under construction; assembly is scheduled to be complete in 2005. However, NASA plans to begin research on the facility as early as 2000, using equipment that has been flown on the shuttle and that can be temporarily installed in modules of the ISS as they are completed. As the ISS grows and more station-specific hardware is ready, the research program will expand and more permanent instrumentation will be fitted into the ISS.

Protein Crystal Growth

A variety of equipment has already been used to grow and observe crystals in space, and innovative hardware continues to be developed today. Having multiple laboratories involved in this process encourages variety and creativity and also prevents NASA from getting locked in to a single hardware approach. However, the efforts of hardware developers need to be coordinated and communications between them must be improved to ensure that different programs are not producing instruments with duplicative capabilities and that technological advances are quickly shared and integrated into all equipment where appropriate.

Recommendation: The efforts of external hardware developers should be coordinated to ensure that instruments are compatible, to prevent duplication of efforts, to ensure that technical innovations are shared, and to facilitate input from the scientific community in defining the goals and capabilities of protein crystal growth equipment for the ISS. NASA must also be prepared to discontinue development projects that do not use cutting-edge technologies or that are out of tune with the most current scientific goals.

A significant factor affecting equipment development is the instability in the budget for the ISS. If money is repeatedly siphoned off from the hardware development work, the equipment on the ISS will be of much lower quality than the cutting-edge hardware available on the ground, and researchers will not be interested in using the outdated equipment or willing to entrust precious samples to it.

The equipment developed by and for NASA should aim to provide a high level of control over samples, equipment, and procedures. On the ISS, crew time will be limited, and the human access to samples and the feedback to the investigators enabled by shuttle trips will be infrequent, so automation and ground-based control of experiments are essential. If principal investigators are able to make decisions about experimental parameters and to adjust experiments in real time, the research results produced in each experiment will be of higher quality, and involvement in the NASA program will be more attractive. Therefore, hardware development efforts should emphasize the importance of automation, monitoring, real-time feedback, telemanagement, and sample recovery (via mounting and freezing).

Effective analysis, preservation, and reentry of promising crystal samples is especially necessary given the key role synchrotrons are playing in protein structure determination. If the NASA program is to attract researchers interested in important and challenging biological problems, ISS hardware must be designed to produce and safely return to Earth crystals of the appropriate size and quality to be analyzed at a synchrotron. However, it is not NASA's responsibility to arrange or guarantee this next step. Building a synchrotron beam line is expensive and would not be the most efficient use of NASA's scarce resources. Assuming that NASA's peer review process is selecting the most scientifically rigorous and interesting projects, successful crystallization should enable researchers to compete effectively for the necessary beam time, and success in this extra layer of peer review should further validate the NASA program within the scientific community.

The X-ray Crystallography Facility (XCF) being designed for the ISS is a multipurpose facility designed to provide for and coordinate all elements of protein crystal growth experiments in space: sample growth, monitoring, mounting, freezing, and X-ray diffraction. The task group was impressed by the XCF, the robotics, the remote control, and the range of experimental capabilities provided. The X-ray diffraction module provides valuable information about whether a given crystal will diffract—this real-time feedback is key to making decisions about the success or failure of a particular crystallization experiment and will help allocate scarce freezer resources by ensuring that the most promising crystals are preserved and returned to Earth.

Finding: Automation, monitoring, real-time feedback, telemanagement, and sample recovery (via mounting and freezing) will be vital for successful protein crystal growth experiments on the ISS. The XCF, through its use of robotics and a variety of experimental and observational capabilities, provides many of the tools researchers need to take full advantage of the microgravity environment.

The XCF is typical of several hardware development projects for NASA in that the technologies it employs can be applied to ground-based research capabilities as well as to those based in space. Currently, however, the scientific community is mostly unaware of the quality of the automation displayed in the prototype of the robotic crystal sample preparation system and of the combined capabilities of the X-ray optics and the low-power source that will be used in the XCF. While commercial entities may need to protect their proprietary work, scientists must have access to full information about all relevant technologies and equipment for the ISS in order to effectively design and execute cutting-edge research in space.

Cell Science

A variety of instruments are being developed to support cell science research on the ISS, including a basic incubator, a perfused stationary culture system, and a rotating-wall perfused vessel (a bioreactor). Overall, the NASA-funded cell science work to date has emphasized the use of bioreactors to support three-dimensional tissue growth. While the development of rotating-wall vessels has had, and should continue to have, a significant impact on cell and tissue culturing methodology on the ground, the task group has a variety of concerns about the effectiveness and appropriateness of this approach for research in the microgravity environment. Issues include the relatively small amounts of data generated per unit volume and the difficulty of accessing the vessel on orbit.

Recommendation: Given the current status of equipment in development, finite fiscal resources at NASA, and the limited amount of volume on the ISS, the task group recommends that future research on the ISS should deemphasize the use of rotating-wall vessel bioreactors, which are already established, and continue to encourage the development of new technologies such as miniaturized culture systems and compact analytical devices.

The final determination on what sort of instrumentation will be most effective for cell and tissue growth in microgravity has yet to be made, and it is important that the relative merits of various pieces of instrumentation be carefully evaluated and that NASA maintain the necessary administrative and engineering flexibility to adopt the most effective systems employing the most advanced technologies and to discontinue hardware development projects that are not attuned to the most current scientific needs of the cell science communities. Close interaction is needed between scientists and the NASA operational personnel responsible for developing and constructing the hardware to ensure maximum flexibility and responsiveness to evolving research goals.

Cellular systems are very sensitive to environmental perturbations. A continuous power supply to maintain appropriate and stable environments during experiments and for sample storage and transport is essential to ensure valid results. A variety of systems are under development to manage power distribution, and care must be taken, particularly during ISS construction, to ensure that cell science experiments are not compromised by power fluctuations. Another issue that will be problematic, particularly during ISS construction but also after the station is complete, is the limited amount of crew time available for research. The automation of routine tasks and ground-based control of experiments will be essential if investigators are to make efficient use of the ISS platform.

Two key supports for automation and ground-based control are (1) sensors to enable physiological control of the cell/tissue culture media environment and (2) analytical equipment to provide feedback about the status of cell and tissue samples. The data from the sensors and the on-orbit analyses should be transmitted electronically in real time to investigators to enable ground-based control of experiments. Scientists on the ground then could select the most important samples for the scarce storage space and could study the changes wrought in samples by freezing and reentry.

Finding: The limited amount of crew time available for research-related work and the infrequency with which investigators will have access to their samples via shuttle trips mean that automation of routine tasks, ground-based control of experiments, on-orbit analytical capabilities, and real-time transmission of digital data are vital for conducting effective cell science research on the ISS.

Refrigeration and freezer capability and transport space are not the only factors limiting the throughput of cell science research on the ISS. Other factors that will affect the size of the program and the number of primary publications include crew time required for the experiments, the amount and reliability of the power supply, adequate storage space and appropriate environments for samples and supplies, shuttle flight schedules to and from the ISS, the volume of materials to be transported, and, of course, the size of the budget provided for cell science hardware development and research support. A window of opportunity has been created by the advances in molecular, cellular, and biochemical approaches (e.g., functional genomics and proteomics) that are occurring as the ISS research platform becomes available. The task group recommends that to most efficiently exploit this opportunity, emphasis should be placed on integration of the different approaches and on collaboration between principal investigators and other researchers inside and outside NASA.

Recommendation: Mechanisms should be developed to enable collaborative research projects that maximize the amount of data obtained from each cell or tissue sample by executing multiple analyses on each sample.

Overall Volume Allotment for Biotechnology Research on the ISS

Currently, NASA plans call for peer-reviewed biotechnology research to occur within one rack on the ISS. This rack would be shared by protein crystal growth and cell science work. In addition, two racks are reserved for the hardware associated with the X-ray Crystallography Facility (XCF) being developed for the NASA Space Product Development Division. The task group considered this arrangement and the needs of the various research communities and recommends a shift in the allotments. Namely, the XCF rack devoted to crystal growth and monitoring should be transferred from Space Product Development to the Microgravity Research Division's protein crystal growth program, where experiments are selected by a centralized peer-review process and a full complement of hardware is available. The rack currently scheduled to be shared by cell science and protein crystal growth can then be dedicated entirely to cell science research.

The task group makes this recommendation based on several considerations. A primary issue is the basic incompatibility between the technical needs of cell science and protein crystal growth equipment on the ISS. The flow of gases and fluids required to maintain rigorous environmental control for cell and tissue culture will produce vibrations that cannot be tolerated by a crystal growth facility. If cell science and protein crystal growth equipment are housed in one rack, one or both of the disciplines will be forced to operate under suboptimal conditions.

The task group also carefully considered the needs of the various research communities expected to use the biotechnology facilities on the ISS. For cell science, there was concern that the amount of data and results generated by half a rack of equipment would not be substantial enough to maintain interest within the scientific community, whereas a full rack's worth of instrumentation could raise the program to a critical threshold. For protein crystal growth, the research community is still uncertain about the benefits of growing crystals in a microgravity environment, so protein sample flight programs are undersubscribed and commercial interest is low. By focusing the protein crystal growth research efforts on biologically challenging problems and by emphasizing hardware capable of monitoring and preserving samples, NASA could direct its resources to validating the program. The current volume commitment of half a rack of general macromolecular research is insufficient to establish the value of the crystal growth program, but a full rack, filled with peer-reviewed experiments that employ all types of available hardware and have access to the capabilities of the XCF, should be adequate to give the program a fair chance of success. If, after several years, the results from the protein crystal growth work have provided sufficient proof of microgravity's benefits and the academic and commercial demand for facilities on the ISS increases, then high-throughput hardware should be developed and the allotment of space on the ISS reconsidered based not only on the demand for macromolecular crystallography research volume but also on the results to that point from the cell science program. Alternatively, if the work done through the augmented commitment recommended here fails to clearly demonstrate the value of microgravity for work on structural biology, then the protein crystal growth program can justifiably be terminated.

Recommendation: The volume allotment for biotechnology work on the ISS should be redistributed as follows:

- *The mounting, freezing, and diffracting equipment of the X-ray Crystallography Facility (XCF) should occupy one rack (as currently planned).*
- *The cell science work should occupy the entirety of what is currently designated the Biotechnology Facility.*
- *The rack currently assigned to the XCF growth equipment and managed by NASA Space Product Development should be officially dedicated to the peer-reviewed macromolecular research run out of the Microgravity Research Division.*

SELECTION AND OUTREACH

NASA research in cell science and protein crystal growth is funded through a collection of approximately 90 active 4-year grants; the total size of the program is roughly $19 million per year. Both ground-based and flight projects are selected through a peer-review process that occurs every other year. While the current grant solicitation mechanism (NASA Research Announcements, or NRAs) is appropriate, it is inadequate to attract the involvement of the best scientists or bioengineers. The task group believes that as the program goes forward, it would benefit from a strengthening of the outreach, selection, and support offered by NASA to ensure that the proposals submitted for consideration are of the highest quality and that everything possible is done to give flight experiments the best chance of success.

Both protein crystal growth scientists and cell science researchers identify themselves with a variety of professional organizations, publications, and conferences, so NRAs should be disseminated to a wider variety of newsletters and announcements in order to reach the multiple communities that might be interested in using NASA biotechnology facilities on the ISS. Another approach to expanding the pool of potential researchers would be to issue NRAs in collaboration with other federal agencies, such as the National Institutes of Health (NIH), the Biotechnology Program in the Engineering Directorate of the National Science Foundation (NSF), the NSF Biological Sciences and Regulatory Biology Divisions, and the Department of Energy. More could also be done to provide sufficient background information for potential investigators who are not familiar with NASA programs. More detail about the special opportunities and constraints of space-based research as well as about the hardware available for the ISS would make it easier for NASA to recruit new applicants for its grants and for those researchers unfamiliar with the NASA program to put together appropriate proposals. Access to information about failed projects would also improve the quality of experiments designed with NRAs in mind and would increase the likelihood of success. In general, results of projects already under way could be more broadly disseminated; however, the task group cautions that presentations should give a balanced portrayal of successes and limitations so as not to raise unrealistic expectations. Misperceptions about the accomplishments of NASA programs can also be gained from press releases that target the general public and portray potential future applications of NASA-funded research as completed or current work. This dissemination of vague or even inaccurate descriptions of its programs seriously diminishes NASA's credibility within the scientific communities.

Recommendation: NASA should improve its outreach activities in order to involve a broader segment of the scientific community in its biotechnology research program and to increase the number of cutting-edge projects submitted for funding. It needs to disseminate NRAs and program results more widely and to provide more complete background information on failed projects and how to design flight experiments.

As the pool of applicants expands, the process of evaluating proposals may also need to be adjusted. NASA's program suffers from longer time scales than are compatible with the current pace of biotechnology research. For example, the 2-year gap between NRA grant submission opportunities is likely to inhibit applications directed at the most cutting-edge research issues. Also, the delay between project selection and flight manifesting of an experiment means that NASA does not always have the hardware flexibility to respond to changes in the field based on new developments in ground-based research (for example, the increased reliance on cryoprotection and freezing of crystals or the use of scaffolding for three-dimensional tissue constructs). Finally, the uncertainties surrounding the NASA budget and the continual schedule changes make people cautious about getting involved in a program that is unable to reliably predict how much money will be available or the schedule for access to the ISS.

One critical step toward raising the profile of the NASA program and the quality of the grant application pool would be to counter the current perception of recipients of NASA funds as a closed community with a fixed membership. On the whole, external input into NASA's priorities for the biotechnology program seems to be relatively limited. Advisory groups are composed of many of the same people who make up the pool of grantees and contribute to the perception that NASA is not really interested in outside input. By reaching out to a broader slice of the protein crystal growth and cell science communities, NASA would not only increase the quality of the advice it receives but would also be able to educate a new group of people about its programs.

According to NASA, the biotechnology Discipline Working Group (DWG) is the main mechanism for receiving advice about the strategic direction of the Microgravity Research Division's biotechnology programs. The group is responsible for providing input to both the protein crystal growth and cell science sides of the program, but in view of the very different scientific objectives and instrumental requirements, having a single working group for these two disparate areas serves no real purpose. If the DWG is split into two groups, each would be able to focus on the issues most relevant to its own scientific area, and the increased number of slots available for each area would give greater breadth to the groups. Care must be taken in selecting new members to ensure that there is not a bias towards those already working with the NASA program. To attract prominent outside researchers to the DWG, the task group suggests that the name be changed to more accurately reflect the group's role as a high-level advisory panel with input on the scope of research announcements, peer review practices, and future programmatic directions.

Recommendation: The separate identities of the protein crystal growth and cell science sections of NASA's biotechnology research program should be emphasized. One key step should be splitting the Discipline Working Group into two strategic advisory committees to reflect the different issues facing each area of research. Prominent scientists not familiar with NASA's programs but aware of the broader issues facing the fields should be recruited to serve on these committees.

An important issue for execution of research in the unforgiving environment of space is the potential for conflict between the scientific goals of an experiment and the engineering limitations associated with a space-based platform like the ISS. Within the biotechnology scientific community, there is the perception that the NASA culture does not emphasize the importance of communication between scientists and operations personnel, nor does it provide tangible assurances to the research community that the execution of high-quality research in hardware designed to answer the most cutting-edge scientific questions is a NASA priority. The community would be reassured by seeing NASA place bioengineers and biological scientists with the appropriate appreciation of research goals and scientifically oriented reflex responses in high enough decision-making positions to ensure that research opportunities are optimally utilized.

Recommendation: The NASA culture tends to limit communication and coordination between operations personnel and researchers during hardware development; between astronauts and investigators before and during experiment execution; and between decision makers and scientists about the allotment of resources in times of crisis. To attract the best investigators to its biotechnology program, NASA must create an environment geared toward maximizing their ability to perform successful experiments.

Protein Crystal Growth

At present, the primary goal of NASA's protein crystal growth program should be to demonstrate microgravity's effect on protein crystal growth and to determine whether studies of macromolecular assemblies with important biological implications will be advanced by use of the microgravity environment. To this end, the task group proposes that NASA instigate a high-profile, nationwide series of grants to support researchers engaging in simultaneous efforts to get both the best possible crystal on the ground and the best possible crystal in space of biologically important macromolecules. The projects funded by these grants should address the uncertainties that have plagued the NASA protein crystal growth program, by using the ISS for a reliable, long-term microgravity environment, by comparing space-grown crystals to the best ground crystals, and by focusing on challenging systems and hot scientific problems. Their results should definitively show whether the use of microgravity can produce crystals of a higher quality than those grown using the best technologies available on Earth. If none of the projects produces a space-grown crystal that enables a breakthrough for the structure determination of a biologically important macromolecular assembly, then NASA should be prepared to terminate its protein crystal growth program. However, if the projects supported by this high-profile, nationwide series of grants succeed in validating the use of crystallization in microgravity to tackle important and challenging problems in biology, demand for the facilities on the ISS can be expected to increase. At that time, NASA should develop an external user program (similar to synchrotron user programs) in which projects are selected by a peer-review committee that includes NASA staff representatives.

Recommendation: NASA should fund a series of high-profile grants to support research that uses microgravity to produce crystals of macromolecular assemblies with important implications for cutting-edge biology problems. The success or failure of these research efforts would definitively resolve the issue of whether the microgravity environment can be a valuable tool for researchers and would determine the future of the NASA protein crystal growth program.

Cell Science

NASA has built a very productive relationship with the NIH based on the development and use of rotating-wall vessels. The NASA/NIH Center for Three-Dimensional Tissue Culture was started in 1994 to expose a wider community to bioreactor technology by allowing researchers from government agencies (e.g., NIH, the Food and Drug Administration, and the Department of the Navy) to test new model systems for biomedical research and basic cell and molecular biology in the rotating-wall vessel hardware with technical assistance from experienced NASA personnel. The task group believes that this outreach program is an excellent idea and recommends that a wider range of investigators be reached by opening the introductory phase of this program to extramural (nongovernment) researchers.

1

Background and Scientific Scope of NASA Programs

INTRODUCTION

The National Aeronautics and Space Administration (NASA) manages research programs in two areas of the rapidly expanding field of biotechnology: protein crystal growth and cell science. The protein crystal growth work focuses on using microgravity to produce higher quality macromolecular crystals for structure determination and on improving understanding of the crystal growth process. The cell science work focuses on basic research that contributes to understanding how the microgravity environment affects the fundamental behavior of cells, particularly in relation to tissue formation and the effects of space exploration on living organisms. The National Research Council's Task Group for the Evaluation of NASA's Biotechnology Facility for the International Space Station was formed to examine and evaluate the use of the International Space Station (ISS) as a platform for research in these two areas.

The logistics associated with space-based research pose a unique challenge to investigators. Strict limitations on the weight and volume of equipment, long planning times, and the remoteness of experiments from the scientists who design them all make research in microgravity—on a space shuttle or on a permanent space platform—very difficult. The reward is an opportunity to work in an environment unavailable on the ground and to examine issues related to gravitational forces, a fundamental part of life on Earth.

The primary motivation for the construction of the ISS has been not the tackling of specific scientific questions but rather the political and cultural need of this country, and perhaps the world, to explore the universe and push at the boundaries of what is humanly achievable. This motivation is not a technical one, and the task group was not set up to evaluate or comment on whether the ISS should be built. Instead, it assumed that assembly of the ISS, having started, would be completed, and focused on how this new research platform could be most effectively utilized. This report therefore discusses whether science in a microgravity environment can advance the fields of macromolecular crystallography and cell science, and if it can, in what areas the advances will most likely occur. The task group discusses issues related to the instrumentation needed to conduct effective research on the ISS and, finally, comments on human factors—how NASA can communicate results and opportunities effectively and enable investigators to design and execute the best possible experiments for the ISS. It offers a variety of recommendations and suggestions for improving the NASA biotechnology research program, and it believes that these changes are necessary if the NASA program is to fulfill the potential for scientific discovery and impact outlined in this report.

PROTEIN CRYSTAL GROWTH

During the 1990s, there was explosive growth in the number and complexity of macromolecular structures being determined by X-ray crystallography, as evidenced by the exponential increase in the number of structures published and submitted to the Protein Data Bank. This growth has been made possible by the convergence of a large number of new technologies, including the following:

- Improved systems for cloning and expressing wild-type and mutant proteins;
- Improved protein and nucleic acid purification techniques;
- Immortalization of crystals by cryogenic freezing;
- Very high brilliance X-ray synchrotron sources;
- Fast, accurate area detectors with high dynamic range;
- Superfast, inexpensive computers; and
- Readily available software packages for data acquisition and reduction, phasing, and refinement.

For the most part, however, protein crystallization is done in much the same trial-and-error manner it was a decade ago, albeit with somewhat less tedium since the introduction of reagent kits and the growing use of automated systems. It is still more art than science. NASA, to its credit, has sponsored a large number of ground-based research projects aimed at understanding the fundamentals of the crystallization process. These included investigations of depletion zones around growing crystals (McPherson et al., 1999), studies of defect formation during protein crystal growth and the effects of these defects on diffraction resolution (Dobrianov et al., 1998, 1999), and analyses of predictors for protein crystallization using light-scattering measurements (Kao et al., 1998; Ansari et al., 1997).[1]

These studies of the crystallization process have occasionally included flight components (McPherson et al., 1999). However, one of the main goals of NASA's program on crystallization in the microgravity environment has been the growth of crystals in space that are of better quality than those available on the ground. In this report, the task group focuses on evaluating the results of the program's effort in this area to date, commenting on the hardware available and in development for future work on the ISS and offering suggestions for improving the project selection process and NASA's outreach to the scientific community.

The Significance of Crystallographic Resolution Limits

The determination of macromolecular structures by X-ray crystallography at a level of detail sufficient for the construction of reliable atomic models requires crystals that diffract X rays to Bragg spacings of 3.5 Å or better. The minimal Bragg spacing to which diffraction measurements can be obtained, loosely referred to as the resolution of the crystallographic analysis, limits the accuracy of the resultant structure in two ways. First, the resolution of the analysis places a limit on the structural features that can be directly visualized in electron density maps calculated using the X-ray data. A resolution of at least 3.5 Å is required to see structural elements in proteins, such as alpha-helices or beta-sheets. Second, once an initial atomic model has been constructed, the resolution of the analysis determines the accuracy with which the parameters of the atomic model can be refined. Positional coordinates in a refined macromolecular structure are determined much more precisely than the resolution of the analysis would indicate. When stereochemical constraints are used in the refinement, such as information about the bond lengths between atoms, the precision of a protein structure typically is approximately 0.5 Å for an analysis carried out at 3 Å resolution and is better than 0.1 Å for a 1.5 Å resolution analysis. In the relatively rare cases where data to better than 1 Å are obtained, individual hydrogen atoms can often be distinguished and the disorder within the protein structure can be described in detail.

---

[1]These are a few examples of the projects under way; a complete list and description of NASA-funded projects in protein crystal growth can be obtained on the Web at <http://peer1.idi.usra.edu/peer_review/taskbook/micro/mg99/mtb.cfm>.

Although a protein crystallographer will always strive to carry out the crystallographic analysis at the highest achievable resolution, the minimal acceptable resolution for a particular crystallographic analysis depends on what questions are being asked. Where the general fold of the protein chain is desired, an analysis at 3.5 Å may suffice to determine the protein structure. However, at this resolution the orientation of hydrogen bonding groups is not well determined, and detailed questions regarding the structural architecture of the protein cannot be answered reliably until a resolution of ~2.5 Å or better is achieved. The precise calculation of the energetics of ligand binding or intermolecular interfaces requires structure determination carried out to an even higher resolution, making possible the mapping of ordered water molecules and an accurate description of hydrogen bonding geometries, and this mandates a resolution of 2.0 Å or better. The most accurate protein structure determinations are carried out at a resolution of 1.5 Å or better.

The intrinsic resolution of a protein crystal can be thought of as arising from two factors. One is the mosaicity, a parameter that is a measure of the misalignment between small coherent blocks of individual molecules within the larger crystal. While crystals that are highly mosaic may diffract to high resolution, the high mosaicity leads to a broadening of the diffraction spots, which can complicate or even foil their measurement. The other crystal characteristic that affects resolution is the Debye-Waller factor, also known as the overall temperature factor, which reflects disorder and mobility within the individual molecules that make up the crystal. Many protein molecules that are of interest today, particularly those that are studied in the form of relatively unstable complexes, are expected to have intrinsically high Debye-Waller factors, limiting the resolution of the resulting diffraction pattern. In such cases, if the size of the perfectly aligned mosaic blocks can be increased, the resulting increase in the sharpness of the diffraction pattern can effectively improve the resolution of the diffraction pattern. In such situations, if growth in the microgravity environment produces crystals with larger mosaic blocks (lower mosaicity), then there may be a significant improvement in the quality of the diffraction measurements. These added levels of detail would enable researchers to see the functional groups and water molecules and thereby more fully understand the interactive mechanisms of macromolecular assemblies.

Today's high-energy synchrotron sources have, in general, eliminated crystal size as the key factor in increasing the diffraction resolution limit. This was not the case when the space crystallization program began, in the mid-1980s. Although the misconception that size is crucial may persist at NASA, scientists today are interested in crystallization methods that provide higher quality crystals, where quality is measured by disorder and mosaicity. Therefore, a well-ordered crystal of average dimensions (around 30 to 50 µm) is all that is needed for effective diffraction studies. Synchrotron technology continues to improve, and the target crystal size may decrease even further before the ISS is completed. Crystal quality, rather than crystal growth, is thus the primary focus of the biological macromolecular research community. The only exceptions are during initial efforts to nucleate protein crystals and when preparing samples for study via neutron scattering.

Goals and History of the NASA Protein Crystal Growth Effort

The current goals of the protein crystal growth efforts funded by NASA's Microgravity Research Division are as follows:

- Understanding the fundamental factors influencing macromolecular nucleation and growth;
- Elucidating which factors may benefit crystal growth in the microgravity environment;
- Growing significantly improved crystals in microgravity for structure determinations;
- Determining the potential of microgravity to solve more complex and challenging crystallization problems, such as integral membrane proteins, glycoproteins, and lipoproteins; and
- Developing technologies and methodologies such as automation and monitoring equipment that would improve the crystallization process on Earth as well as in space.

The program began with exploratory efforts to grow macromolecular crystals in space in 1985. From that time through October 1999, experiments relating to the crystallization of biological samples were carried out on 43 NASA missions. This has resulted in a total of 185 different proteins and other biological macromolecular assemblies being studied. Overall, the results from the program so far are inconclusive. In many of the cases that have been listed as successful, the improvements obtained were incremental advances in resolution rather than substantial increases. In certain cases, such as the result reported for $T^3R_3^f$ insulin, where a sizeable improvement in resolution appears to have been obtained by the application of microgravity (Smith et al., 1996), the impact of the work is lessened by the fact that a considerable amount of structural information was already available for insulin. Consequently, the enhanced resolution observed for the space-grown crystals did not enable a distinctly new body of information to be obtained. Due to both the limited number of experiments and the type of proteins for which significant improvements were noted, the impact of microgravity crystallization on structural biology as a whole has been extremely limited. At this time, one cannot point to a single case where space-based crystallization efforts produced a crucial discovery leading to a landmark scientific result. In addition, the difficulty of mounting simultaneous efforts to produce the best possible crystals both on the ground and in space has limited the ability of researchers to make the comparisons between microgravity and Earth crystals that would be necessary to demonstrate that the microgravity environment can produce superior crystals.

Finding: The results from the collection of experiments performed on microgravity's effect on protein crystal growth are inconclusive. The improvements in crystal quality that have been observed are often only incremental, and the difficulty of producing the appropriate controls limit investigators' ability to definitively assess if improvements can be reliably credited to the microgravity environment. To date, the impact of microgravity crystallization on structural biology as a whole has been extremely limited.

Despite the lack of impact of microgravity research on structural biology up to now, there is reason to believe that the potential exists for crystallization in the microgravity environment to contribute to future advances in structure determination. All research on protein crystallization in space so far has been done under suboptimal conditions. Most of the work has been done on fairly short space shuttle flights, with a few experiments occurring on the Russian Mir space station. The crystallization work on the space shuttle has been restricted to a matter of days, which is not enough time in most cases to complete the crystallization process, especially in space, where crystals appear to nucleate and grow more slowly. Except for space shuttle missions devoted exclusively to microgravity research, the environment on the space shuttle has not been noise- and vibration-free. No mechanism has been provided to stabilize the crystals that do grow and to protect them from the stresses of reentry. In general, the ability to visualize crystal growth in space has been extremely limited, preventing investigators from determining if flawed crystals examined after landing had failed to grow well in space or if crystals with good morphology had indeed been grown but later had been damaged during reentry. The irregular schedules of shuttle missions and the long lead times have made it difficult for scientists engaged in extremely competitive structural analyses to seriously consider participation in the shuttle-based crystallization experiments. Long delays between shuttle flights has meant that lessons learned from one flight are often not translated into improved experiments on a subsequent flight. The slow and uncertain progression of experiments on the space shuttle has disconnected them from the even more rapid tempo of contemporary protein crystallography research.

Results to Date:
Examples of Successful Experiments and the Importance of Defining Controls

Despite these limitations, it is conservatively estimated that for 36 of the 185 different proteins and other biological macromolecular assemblies that have been studied in space, the resolution of the crystallographic analysis was better than that of the best ground-based results available at the time. The proteins whose resolution improved in space range from well-understood test cases, such as lysozyme, to proteins that present significant challenges for contemporary structural biology, such as the EcoRI-DNA complex, the nucleosome core particle, and the epidermal growth factor receptor. Enhanced resolution has also been obtained for proteins of importance for drug design, including the HIV protease complexed with lead compounds and the influenza neuraminidase.

While this list of improved crystals is certainly tantalizing, it is difficult to draw concrete conclusions from the limited data available, for the reasons discussed in the previous section. Below, the task group describes in detail four of the positive results. These cases were selected because the researchers were able to cleanly compare crystals grown in space to the best crystals produced using ground-based systems and because the improvements in resolution were substantial. The four examples are lysozyme, canavalin, satellite tobacco mosaic virus (STMV), and insulin. While the proteins in these experiments are not in themselves necessarily of great biological significance, the studies are important because of what they indicate about the potential benefits of crystal growth in space.

Lysozyme, a workhorse in the field of protein crystal growth, yields crystals in space that have much better properties than those of crystals grown on Earth. Analysis of tetragonal lysozyme crystals grown on two space shuttle missions showed, impressively, that the mosaicity of the crystals was improved by factors of 3 or 4 over that observed for lysozyme crystals grown on Earth (Snell et al., 1995). The observed reduction in mosaicity is a very significant improvement, because it can allow the measurement of very weak reflections that would otherwise be too broad to be observed over background. Although crystals of lysozyme with very low mosaicity can occasionally be obtained on Earth, only about 1 in 40 of them have properties comparable to those of the crystals grown in space.

STMV is a small icosahedral plant virus, consisting of a protein shell made up of 60 identical protein subunits of molecular weight 14,000. The crystallization of STMV has been studied extensively on Earth. Its crystallization in microgravity was investigated during two space shuttle missions, in 1992 and 1994. Using a liquid-liquid diffusion technique with careful temperature control (in an experimental setup known as CRYOSTAT), remarkably large crystals of STMV were obtained (Day and McPherson, 1992). In most cases the crystallization chambers contained large single orthorhombic crystals, the largest of which measured 1.5 mm in length and 1.0 mm in both of the other directions. These crystals are about 10 times larger in volume than the largest crystals of STMV previously grown in ground-based laboratories. Particularly noteworthy is the fact that, in contrast to the crystals grown on Earth, the crystals grown under microgravity conditions were visually perfect, with no striations or clumping of crystals. Furthermore, the X-ray diffraction data obtained from the space-grown crystals was of a much higher quality than the best data available at that time from ground-based crystals. The average intensity of the diffraction measurements relative to the standard deviation was seen to be increased substantially over the entire resolution range, resulting in nearly 50 percent more X-ray data than had previously been available. STMV also crystallizes on Earth in a cubic crystal form that diffracts poorly; at the time of the 1994 shuttle flight, the best available ground results gave only about 6 Å resolution (Koszelak et al., 1995). Cubic crystals of STMV obtained on board the space shuttle were 30 times larger than those obtained on Earth. These crystals diffracted X rays to 4 Å resolution, a significant improvement over the ground-based crystals.

Canavalin is a plant storage protein from the Jack Bean and is a trimer of three identical subunits of molecular weight 47,000. It can be crystallized reliably on Earth, which has made it one of the proteins commonly used for crystallization studies. As with STMV, large crystals of canavalin were obtained in space (Koszelak et al., 1995). Visually perfect rhombohedral and hexagonal crystals of canavalin with edges >1 mm in length were obtained in large numbers, with significantly better diffraction properties than those of crystals grown on Earth. For the rhombohedral crystals, the diffraction limit was extended from 2.6 Å to better than 2.3 Å. The improvement in resolution for the hexagonal crystals was more impressive: it went from about 2.7 Å (Earth) to nearly 2.2 Å (space). In both cases the total number of useful X-ray measurements essentially doubled.

The crucial human hormone known as insulin consists of two chains, an A-chain consisting of 21 residues and a B-chain consisting of 30 residues. Insulin aggregates to form hexamers, which undergo allosteric transitions between R and T states. The switching between the R and T states is altered by the presence of particular ions and organic molecules, and there is interest in identifying additives that would stabilize the R-state over the T-state, which would lead to insulin preparations with greater stability. To this end, high-resolution crystallographic analyses of insulin are being carried out. This project has led to a particularly clean comparison between the results of ground-based and space-based crystallizations of a protein. Human $T^3R_3^f$ insulin obtained from a single source was crystallized on Earth and also in microgravity using batch crystallization (Smith et al., 1996). The crystals grown in space were larger and free of imperfections compared with crystals grown on Earth. Strikingly, whereas data to 1.9 Å resolution were obtained using the crystals grown on Earth and a laboratory X-ray source, data to 1.4 Å resolution were obtained using the space-grown crystals and the same X-ray source. In follow-up work, preliminary results indicate that data to 0.9 Å resolution have now been obtained using synchrotron X-ray radiation on T^6 insulin crystallized on the space shuttle in 1998 (G.D. Smith, personal communication). This ultrahigh resolution data is allowing very detailed analysis of the molecular structure, including the study of electronic distributions within the protein molecule.

The four cases described above provide the most convincing data currently available on the benefits of growing protein crystals in the microgravity environment. The 32 other experiments that produced space-grown crystals with improved resolution also support the potential value of microgravity. However, in some of these other cases, the investigators were not able to make the comparisons needed to demonstrate that growth in the microgravity environment was indeed the factor responsible for producing higher quality crystals. It is difficult to carry out a completely controlled experiment, particularly for cutting-edge projects that are intrinsically difficult and that involve structure determinations of immediate interest to researchers. It is not enough to compare space-grown crystals to crystals grown on Earth in the same equipment and solution over the same time period; the microgravity-grown crystals must also be compared to the best result from all Earth-based attempts at growing the crystal regardless of crystallization conditions, equipment, or time of growth. This latter comparison is the baseline standard for defining success.

The complexities that arise when trying to make appropriate comparisons are exemplified by in the case of the restriction endonuclease EcoRI complexed with DNA (EcoRI-DNA). High-quality crystals for this assembly have been obtained in microgravity (J. Rosenberg, 1999, submitted to *Proteins*), and the diffraction data obtained from the space-grown crystals are of significantly better quality than those obtained from similar samples on Earth. This is of considerable interest since the study of protein-DNA complexes is often plagued by crystals of poor quality. While this study demonstrates that biologically important results can be obtained from protein crystallization in space, it turns out that the incorporation of additional features in the analysis of the space-grown crystals, such as the use of cryogenic techniques and synchrotron radiation, makes it difficult to be certain that the improvements are due to microgravity and not to some of these additional factors (J. Rosenberg, personal communication). A direct quote from Dr. Rosenberg nicely summarizes the problem: "Significant improvement in the resolution of our EcoRI-DNA cocrystals . . . was due to a number of factors, including microgravity . . . [H]owever, I'm not in a position to untangle all the factors and just don't have the data to say how much of the improvement was due to each of the factors."

Potential Areas of Future Impact

There is now a certain amount of evidence that crystal growth in a microgravity environment can have beneficial effects on the size and intrinsic order of macromolecular crystals. In many cases, crystals obtained in space are larger, have lower mosaicity, and diffract to higher resolution than comparable crystals grown on Earth. However, space-based crystallization programs have been very limited in scope in terms of the total throughput of samples compared with the enormous reach of modern protein crystallography on Earth. In addition, space-based crystallization efforts have been carried out under extremely adverse conditions. Therefore the results of the program, while intriguing, have had an extremely limited impact on biology during a time when technological innovations on the ground have produced significant and fundamental advances in our understanding of protein behavior and interactions.

However, despite the greatly increased sophistication of ground-based protein crystallization projects, the crystals of many important targets have suboptimal diffraction characteristics. Improvements in diffraction that move a system from the margins of structure determination (3.5 to 3.0 Å) to well beyond that boundary will have a significant impact on the ability of the resulting structure to provide important insights into biological mechanisms.

An example of such a situation is provided by the potassium channel. Potassium channels are integral membrane proteins that are important elements in the functioning of neuronal cells, and they also play diverse roles in the physiology of many different cell types. The potassium channels of greatest interest are those found in mammalian, particularly human, cells. However, it has not yet been possible to obtain crystals of mammalian potassium channels that are suitable for X-ray crystallographic analysis. Instead, crystals have been obtained of a bacterial homolog that is similar in structure to the central core of mammalian potassium channels. While this analysis has provided a structural model for potassium channels and has revealed the general features of ion conductance and selectivity, diffraction data from these crystals are very weak and anisotropic at better than 3.5 Å. There would be enormous value in improving the structural accuracy of the model for potassium channels, but despite very significant efforts, better-diffracting crystals have not yet been obtained (R. MacKinnon, personal communication).

If the protein or proteins being crystallized are soluble, relatively stable, and well defined in their conformational state, there is little doubt that extensive experimental manipulation in the laboratory will eventually lead to better-diffracting crystals. For membrane proteins, such as the potassium channel, the difficulties appear to be much more serious. One reason for this is that the hydrophobic interactions that stabilize membrane proteins within the lipid bilayer are relatively nonspecific compared with the hydrogen bonding interactions that occur between surface side chains in soluble proteins. This makes it very difficult to obtain membrane protein crystals that diffract to high resolution, so membrane proteins are attractive targets for investigation in microgravity environments. Another general class of proteins yielding crystals that diffract very poorly are those that form transient complexes during dynamic events, such as during cellular signaling. There is great interest in obtaining high-resolution structural analyses of such protein complexes, and these may benefit from the particular conditions of microgravity. Drug design projects are another case where microgravity may be important. In the design of inhibitors it is usually important to see the stereochemistry by which binding occurs, and it is also necessary that the crystal structure be obtained for the precise target in question rather than for a closely related protein. This is a restriction that is usually avoided in practice, since the protein crystallographer will often search a set of closely related proteins for a protein with optimal crystallization characteristics. It is not at all uncommon to find that the particular protein that is most interesting, for example, the human variant of a family of proteins, does not yield suitable crystals.

The relatively poor diffraction obtained for such systems can arise for one or more reasons. These include the intrinsic flexibility of the macromolecular system being crystallized, as well as impurities or other factors that impede optimal crystal growth. At present there is no direct information on whether crystallization in a microgravity environment will have a positive impact in cases where the sole inhibitor of crystallization is the intrinsic flexibility of the molecules involved. Further experimentation will help resolve this question, but the controlled manner in which crystals grow in a microgravity environment may be beneficial in these cases.

Finding: While enormous strides have been made in protein crystallization in the last decade, it is still the case that there are very important classes of compelling biological problems where the difficulty of obtaining crystals that diffract to high resolution remains the chief barrier to structural analysis of the crystals. It is here that the NASA program must look to maximize its impact.

A prerequisite for all macromolecular structure studies is the availability of crystals that have suitable morphology and that are well ordered, sufficiently large, and stable enough to permit the recording of high-resolution diffraction data. Crystallographers will beat a path to any technology that can provide better quality crystals. Obtaining large crystals is probably not as important as it once was, because most crystallographers have access to very high brilliance X rays at synchrotron sources and can work with quite small samples. Similarly, cryogenic freezing techniques have been perfected to the point where most crystals can be kept stable indefinitely.

The main potential advantage of microgravity, therefore, is the possibility of obtaining crystals that diffract to higher resolution or crystals that have more favorable morphology. This could be especially important in structure-based drug design. Since the number of observable diffraction maxima increases as the inverse cube of the resolution, a crystal with a resolution of 2.0 Å yields nearly twice as many data as a crystal with a resolution of 2.5 Å. Determining the orientation of a small molecule inhibitor in an electron density map calculated from 2.5 Å data is problematic at best, whereas at 2.0 Å it will often show up quite clearly.

Potential Benefits of the Space Station Platform

Use of the ISS for future microgravity crystallization projects will probably lead to dramatic changes in the NASA macromolecular crystallography program. The length of experiments will increase significantly. Regular shuttle flights to and from the ISS will allow for considered planning of crystallization experiments. Improvement and standardization of the crystallization hardware will allow laboratory scientists to optimize crystallization procedures for the specialized hardware, maximizing the chances for success. The incorporation of microscopic examination on board the space station, a crucial element of successful crystallization in space, means that the crystallization process can be monitored and successful crystallization recognized when it occurs. The coupling of microscopic examination with automated procedures for crystal recovery and freezing will dramatically improve the ability of scientists to bring back high-quality crystals from space.

The benefits that structural biologists will realize as their use of the space shuttle on an ad hoc basis is replaced by the deployment of a dedicated protein crystal growth facility on the ISS can be seen as parallel to the benefits they realized when they started using synchrotron facilities dedicated to the production of X rays (see Box 1.1). For microgravity crystallization, the transition to a much more predictable and ordered regime on the ISS will have a maximum impact on modern biology if the projects chosen for experiments are ones that require improved crystallization to achieve significant scientific breakthroughs.

BOX 1.1 Analogy Between Synchrotrons and Space-based Research Platforms

There is a striking and instructive analogy between the development of synchrotron X-ray sources and the development of microgravity crystal growth facilities in the United States. Both rely on "big machines" and large capital expenditures of government money. The results of early experiments on both did not seem to many crystallographers to justify the expenditure.

First-generation synchrotrons, such as those at Cornell and Stanford Universities, were built for high-energy physics research. X rays were available for macromolecular diffraction studies only occasionally, and then only on a parasitic basis. While early experiments yielded some useful results, most investigators found these facilities difficult and frustrating to use. Often, the meager results obtained did not justify the time and money spent. In the 1970s and early 1980s, few of the early investigators could claim that synchrotron sources had had an important impact on structural biology. In 1982, the National Synchrotron Light Source, a second-generation machine dedicated to the production of X-rays, came on line at least 2 years behind schedule. Beam time on this machine was scarce, and most beam lines were not dedicated to macromolecular diffraction. However, early experiments soon proved the worth of synchrotron sources to all but the most skeptical.

This led to the development of third-generation machines such as the European Synchrotron Research Facility, the Advanced Photon Source, the Advanced Light Source, and Spring 8, with greatly increased X-ray brilliance, large numbers of user-friendly beam lines, and associated laboratories. These facilities have now become the X-ray sources of choice for most crystallographers.

By analogy, the space shuttle has been the first-generation microgravity platform. It was built as a space training and transport facility. Most microgravity experiments were consigned to small middeck lockers in the crew's living quarters. Generally, fewer than a hundred crystallizations per flight were attempted, and most were allowed to run for only a week or less. Although there have been some intriguing successes from the experiments carried out to date, at least as many crystals were lost before they could be returned to Earth-based laboratories for study. Few would claim that the program has yet had a significant impact on structural biology.

The International Space Station (ISS) may be considered the second-generation platform. Experiments will be carried out in dedicated racks in the ISS modules. Many more crystallizations will be set up and allowed to proceed for weeks or months, with periodic visual monitoring both on the ISS and from the ground. In addition, it may be possible to automate the process of crystal growth, monitoring, mounting, and freezing, and of obtaining diffraction data in microgravity owing to recent technological advances in hardware, especially the instrumentation being developed by scientists and engineers at the Center for Macromolecular Crystallography in Birmingham, Alabama.

It was the second generation of synchrotrons that demonstrated to the scientific community at large their potential value to structural biologists, and the ISS has the potential to produce the data necessary to resolve the worth of microgravity crystallization in a definitive fashion.

In order to engage the research community, NASA must focus its support on programs that are developing technologically innovative equipment and engaging in the structure determination of crystals with important biological implications. While past NASA-supported research on the crystallization process has not been without value, NASA's priority should now be to resolve questions about the usefulness of protein crystal growth in the microgravity environment to tackle important biological questions. Until the uncertainty about the value of space-based crystallization is resolved, a program of this fiscal magnitude is bound to engender resentment in the scientific community.

Potential for Interest from Commercial Entities

The focus of the task group's charge was to evaluate NASA's biotechnology program and facilities for the ISS. However, the task group also considered the related question of what role industry might play in developing what will be a large international facility for protein crystal growth. Commercial users of a macromolecular crystal growth facility on the ISS would come almost exclusively from pharmaceutical and biotechnology companies, with perhaps an occasional user from a contract research organization or an instrument manufacturer. Worldwide there are currently more than 70 companies with research programs in macromolecular crystallography. In aggregate these industrial organizations employ approximately 300 scientists and technicians with all levels of expertise in crystallography. Most are located in countries that already participate in development of the ISS.

All of these companies employ crystallographers to aid in the design of biologically active molecules for use in human and animal health care or agriculture for the production of food and fiber. Industrial research programs in macromolecular crystallography fall within two broad categories. In structure-based drug design, the three-dimensional structure of a target macromolecule is determined to help in the design of a compound, most often a small molecule, that will bind tightly and selectively to the target, modifying its activity. In macromolecular engineering, the structure of a macromolecule is determined in order to guide research aimed at changing its structure. The goal is to alter its properties in some desirable way, with the final commercial product being the mutant macromolecule itself.

Although many pharmaceutical and biotechnology companies have participated in the microgravity crystallization research, not one has yet committed substantial financial resources to the program. This is likely to remain the case until the benefits of microgravity can be convincingly documented by basic researchers and until facilities in space can handle greatly increased numbers of samples in a much more user friendly manner. The future financial participation of industry is, however, not out of the question. Again, the analogy of synchrotron beam lines is appropriate. About a decade ago, 12 of the largest pharmaceutical companies doing macromolecular structure research in the United States formed a consortium to build beam lines at the Advanced Photon Source.[2] To date these companies have invested approximately $10 million in what appears to be a very successful venture.

CELL SCIENCE

Goals and Potential Impacts of the NASA Cell Science Effort

The mission of the cell science program in NASA's Microgravity Research Division is to obtain new knowledge and increase the understanding of how low gravity influences fundamental cell biology with respect to tissue formation and space exploration. A variety of factors motivate this work. A primary goal is the need to understand the potential impact of the microgravity environment on the cell, tissue, and organ system functions of astronauts

---

[2]More information about the Industrial Macromolecular Crystallography Association and its 12 member companies is available on the Web at <http://www.imca.aps.anl.gov/>.

spending months in space and on biologically based life-support systems such as those for plant growth or waste treatment. Cell science investigations on the ISS may help to foresee problems encountered by future longer-range space travelers. The research also has implications for ground-based systems, as perturbations of biological systems by microgravity can provide insight into physiological control in the absence of mechanical forces and in the absence of convection. This work could give scientists insight into how cellular processes respond to mechanical and chemical manipulation, eventually allowing them to design more efficient bioprocesses and to develop a new generation of high-resolution biosensors. Finally, the program also allows investigators to compare various tissue culturing techniques to determine which of them produce systems that most effectively mimic the characteristics of genuine tissue.

NASA's program on cell science in the microgravity environment is fairly young.[3] As researchers gained experience the goals of the program broadened and the potential impact of the work became better understood. Originally, NASA focused on the generation of three-dimensional tissue constructs and on the rough characterization and comparison of these constructs to natural tissues, envisaging the commercial exploitation of space for generating large amounts of tissue. The task group does not believe such a goal is realistic and is encouraged to note that recent NASA-funded work has focused on more basic research. In the long-term, NASA's work in cellular biotechnology is aimed at proof of concept and the development of a research platform for external investigators. According to NASA, its criterion for success is having the NASA-funded work of today lay the groundwork for a big breakthrough 10 years from now. In what area such a breakthrough will occur is not easy to predict, but a wide-ranging program on cell science in microgravity has the potential to impact a number of areas. In the broadest of terms, these areas include an increased understanding of the basic cell biology of terrestrial life in space and the effects of gravity on basic cell biology on Earth; the production of biopharmaceuticals and of functional tissue constructs for research and medicine; the propagation of organisms for antibiotic and vaccine development; and the identification of technologies that will result in new products to advance scientific capabilities on Earth. More specific areas of impact might include the use of bioreactors for efficient high-fidelity production of complex proteins requiring significant post-translational processing; the propagation of parasites for evaluation of function; the propagation of tumor tissue for evaluation of responses to therapeutic options; the miniaturization of analytical equipment such as flow cytometers, mass spectrometers, and sensing systems; a better understanding of gene expression within the three-dimensional context of cell and tissue architecture; and an appreciation for the consequences of modified gravitational forces during launch, sustained periods in space, and reentry. Because many aspects of cell behavior and tissue growth under low gravity conditions are not well understood, the basic cell science research carried out on the ISS will increase the amount of information available to the community. Whether this new knowledge about cells and tissues in different environments will fundamentally alter the scientific understanding of biological functionalities remains to be seen.

The enormous range of specific biological questions that fall within the general goals of the NASA cell science program is both an opportunity and a burden. On the one hand, since it is difficult to say which specific area has the greatest potential for a significant discovery, it could be a mistake to eliminate entire branches of research from contention for NASA funding. On the other hand, resources available on the ISS, such as equipment volume and crew time, will be limited, and in order to most effectively exploit this new research platform, NASA, in consultation with a committee representing the cell biology research community, might want to more narrowly define a subset of areas in which to support investigations. A clearer statement of goals would allow instrumentation developers to focus on specific needs for culture environments and analytical equipment on the ISS. Also, if focused on a few specific program areas, ground-based NASA-sponsored research projects might be more likely to construct the body of knowledge that would enable the experiments on the ISS to produce the future big breakthrough sought by NASA.

---

[3] A small amount of funding (less than $1.5 million per year) was allotted to cellular biotechnology work within the Microgravity Research Division from 1983 to 1992, but significant resources (over $5 million per year) were not devoted to the program until the 1993 fiscal year.

Finding: It is appropriate for NASA to support a cell science program aimed at exploring the fundamental effects of the microgravity environment on biological systems at the cellular level. Results from such basic research experiments could have a significant impact on the fields of cell science and tissue engineering. However, the specific important questions within cell biology that can best be tackled on the ISS do not seem to have been defined yet. Narrowing the broad sweep of the current program might focus instrument development efforts and accelerate progress toward complete understanding of the effects of microgravity on specific biological phenomena.

The scope of NASA-supported cell science research is broadening. The research now goes beyond evaluating a limited number of structural and functional parameters and is taking a more mechanistic approach, using gene expression measurements, for example. It is also moving beyond a focus on instrument development to investigating the science that can be achieved using the new equipment. Detailed analyses at the biochemical and genetic level have demonstrated that microgravity has statistically significant effects on fundamental biological processes such as signal transduction and gene expression. Thus, microgravity, as an experimental parameter, may provide insight into fundamental aspects of biological regulation that will be important in terrestrial as well as extraterrestrial environments. Further, among systems tested to date, tissue constructs grown in microgravity have shown a unique utility for supporting studies on viral and pathogen culture. Key areas in which perturbations of cell structure and function in the extraterrestrial environment might be observed are components of nuclear architecture, cytoarchitecture, and the extracellular matrix. It is becoming increasingly evident that the organization of genes and regulatory proteins within the nucleus, the organization of nucleic acids and signaling proteins in the cytoplasm and cytoskeleton, and the organization of regulatory macromolecules within the extracellular matrix contribute to the physiologically responsive fidelity of gene expression. Consequently, the functional interrelationships between cell structure and gene expression within the three-dimensional context of cell and tissue organization should be rigorously and systematically studied under microgravity and regular Earth-gravity conditions. The corollary is that microgravity conditions can provide valuable insight into structure-function interrelationships that connect control of gene expression to cell and tissue architecture. Regulatory events within the three-dimensional context of cell and tissue organization should be further pursued under microgravity conditions and compared with similar constructs under controlled conditions. The information gained in these sorts of studies will certainly be useful, but it is important to note that, given the large number of signaling molecules and genes that could be investigated, a specific theme or focal point to the NASA program would improve the prospects of fully characterizing any particular cell structure-gene expression interrelationship.

To date, NASA's work in cell science has taken place on shuttle flights and on the Mir space station. These experiments, and investigations using other space facilities, have demonstrated that microgravity and the space environment affect cell shape, signal transduction, replication and proliferation, gene expression, apoptosis, and synthesis and orientation of intracellular and extracellular macromolecules (Dickson, 1991; Moore and Cogoli, 1996; Cogoli and Cogoli-Greuter, 1997; Lewis et al., 1997; Freed et al., 1997; Hammond et al., 1999). With the increased availability of research opportunities on the ISS and the new hardware developed specifically for this platform, further investigation of these processes may clarify how cells behave in the microgravity environment. For example, a deeper comprehension could be sought about the mechanisms behind the observed effects of microgravity on cells relevant to human physiology (e.g., muscle, bone, balance, circulation); understanding of how the cells in these systems sense and respond to mechanical stresses (i.e., the absence of gravitational acceleration associated with microgravity) would have immediate relevance for NASA's manned space program.

Potential research topics would not be limited to areas that have already been explored, but could come in other areas, including the adaptive responses of cells in microgravity to factors such as radiation; induced phenotypic and genotypic changes; selective pressure of the space environment on replicating cells; and the effect of microgravity on plant cells and tissues, on microorganisms (e.g., those that cause disease or that will be used for sewage treatment on long-range flights), and on cells (e.g., osteoblasts) that may not proliferate in bioreactors as they are currently designed. More general areas of study might include bioreporter models and sensors for biomolecular signatures and propagation of obligate and facultative parasites.

A key element in NASA's program in the future should be a concerted effort to understand the artifactual effects of the microgravity environment on cell science experiments (NRC, 1998). That cell and tissue constructs behave differently in space has been established; experiments designed to pinpoint the causes of the changes are the logical next step. In microgravity, the fluids that make up the culture environments behave quite differently than on Earth, and the differences in mass transport, convection, and buoyancy-driven flows could have a variety of effects on cells and tissue that are not directly related to the absence of gravity. While it may be reasonable to expect gravisensing cells to significantly alter their behavior in space, explaining the response of non-gravisensing tissue grown in the microgravity environment is unlikely to be straightforward. Efforts to separate the effects of low gravity on cells from the effects of low gravity on the cell growth medium will require meticulous experimental design, quantitative measures of cell alteration, and careful investigation of multiple experimental control groups. A thorough exploration of the factors affecting cell behavior in space will not only increase understanding of the effects of microgravity on cell and tissue formations but may also reveal unexpected information about the basic interactions between cells and their growth media.

Experimental Design and Instrumentation

The cell science program encompasses a wide range of research topics, from cancer cells to parasites, from chondrocytes to Bowhead whale cells. Many of the ongoing projects are thematically linked by a focus on three-dimensional tissue formation, and some success in this area has been achieved both in space and in ground-based experiments. However, tremendous progress has also been made in three-dimensional tissue development on Earth under unit gravity, using, for example, scaffolds and extracellular matrix gels. In experiments done in space, cell cultures experience a different gravitational environment, which reduces convection, buoyancy-driven flows, and sedimentation, and it is difficult to separate the various factors causing differences between space- and Earth-grown samples, making it difficult, in turn, to determine the appropriate experimental controls for space research. Possible approaches include the use of bioreactors on Earth, culture bags in the microgravity environment, bioreactors in space, three-dimensional structures grown on Earth from scaffolds, and the same experimental setup operated in the on-orbit 2.5-m centrifuge to restore the effects of unit gravity. The task group believes that for each experiment done in space, several potential control groups must be evaluated. Ordinary tissue culture flasks or spinner flasks on Earth are not appropriate benchmarks.

A related issue is the need to develop and apply quantitative measures for evaluating the results of cell science experiments performed on the ISS. In the space shuttle-based research to date, technical limitations, such as unstable environments and reentry effects, have produced experimental results that, while provocative, were essentially descriptive and phenomenological. Measurements are needed that can illuminate the molecular mechanisms underlying cellular functions. Work on techniques such as gene expression enables a more detailed accounting of what has been observed and can focus future research on areas where the effects of microgravity are greatest. Quantitative approaches may also help researchers to distinguish among the array of factors that may be influencing cell and tissue characteristics in the microgravitiy environment. Currently, researchers are limited by the difficulties inherent in distinguishing the effects of launch, flight, and, in many cases, reentry on samples. Experiments should be designed to separate these effects; such experiments, and the careful interpretation of the resulting data, will require close and frequent interactions with investigators in NASA's Life Sciences Division, as discussed below. A better understanding of these effects, as well as the development of quantitative approaches, will also assist in determining whether bioreactors, which were originally developed to simulate the microgravity environment, provide appropriate predictions of the behavior of cells and tissue in such an environment. The answer to this question will not be known until comparisons are made with experiments that have been subjected to microgravity environments and not modified by launch and reentry.

Finding: Appropriate experimental controls for space-based cell science experiments have not yet been determined. The best controls would be those that enable researchers to separate and investigate the multiple factors— including launch and reentry, effects of microgravity on the culture medium, and direct effects of microgravity on cellular behavior—that produce the changes observed in cells and tissues grown in space. Analytical techniques that measure the molecular mechanisms underlying cellular functions will be essential to provide data for comparing proposed experimental controls and quantifying the observed changes in cell and tissue samples.

Requirements for Interprogrammatic Coordination Within NASA

At NASA, the work viewed by the task group was being carried out in the biotechnology section of the Microgravity Research Division. The themes of the cell science work under way in this program overlap with the scope of the work ongoing in the NASA Life Sciences Division, such as research on bone formation and muscle function. The complementary nature of these two programs needs to be recognized so that NASA personnel and external researchers can take full advantage of the resulting synergies. The Life Sciences Division covers a broad array of topics, including cellular and molecular biology, gravitational ecology, and organismal and comparative biology, that potentially relate to the cell science work under way in the Microgravity Research Division. Clearly, NASA and NASA-sponsored researchers would benefit from sharing and coordinating experiments on similar cell biology projects, such as the work on muscle growth and on osteoblasts. There is also a potential for synergy in connection with the Life Sciences Division's work on larger systems. Through this research, the observations of whole organisms provide a basis for theories about what happens to cells and tissues in microgravity, while the Microgravity Research Division's cell science program greatly expands the range of hypotheses that can be tested. Cooperation between the two programs provides a way to relate observations on cellular constructs to whole-animal response, for example, whether gene expression patterns in microgravity tissue constructs are similar to gene expression patterns in the corresponding animal organ in response to microgravity. If NASA-sponsored research is to provide a cellular basis for understanding the physiological effects of prolonged weightlessness on astronauts, a more complete continuum between the cellular experiments and whole organism responses needs to be established, and nonoverlapping but related contributions from both the Life Sciences and Microgravity Research Divisions are necessary.

Coordination (not just communication) between the divisions on cell science work is needed for fully understanding the relationships between in vivo and in vitro systems, and a better sharing of resources and expertise seems essential. While there is already overlap in flight hardware availability (both life science and microgravity researchers have and will continue to have access to the same equipment), it is not possible to have projects that are jointly funded by the Life Sciences Division and the cell science section of the Microgravity Research Division. The task group believes that although the potential synergies are significant, they are not yet being realized. A mechanism to establish cosponsored projects should be considered, possibly via joint NASA research announcements. It is important to note that each program does have unique approaches, goals, and perspectives, and to maintain these valuable differences, the administrative integrity of the separate programs should be retained.

Recommendation: The research strategies and projects of the cell science work in the biotechnology section of the Microgravity Research Division should be more closely coordinated with the work of NASA's Life Sciences Division to take advantage of overlapping work on bone and muscle constructs and of potential synergies between in vitro and in vivo research projects.

2

Instrumentation

LOGISTICS FOR USING THE INTERNATIONAL SPACE STATION AS A BIOTECHNOLOGY RESEARCH PLATFORM

The International Space Station (ISS) is currently under construction; assembly is scheduled to be complete in 2005. However, NASA plans to begin research on the facility as early as 2000, using equipment that has been flown on the shuttle and that can be temporarily installed in modules of the ISS as they are completed. As the ISS grows and more station-specific hardware is ready, the research program will expand and more permanent instrumentation will be fitted into the ISS. The present schedule calls for a specialized biotechnology facility to be one of the last units installed on the ISS in 2005, so until that date, the hardware for protein crystal growth and cell science research in space will be fitted into EXPRESS racks in whatever laboratory modules have been completed.[1] A more detailed outline of the schedule for research on the ISS is included in Appendix A.

Although equipment from the space shuttle will be used, temporarily, on the ISS, there are several key differences between the logistics of experiments on the shuttle and on the ISS. First are the time scales. On the shuttle, experiments lasted no more than 2 weeks at a stretch. On the ISS, the microgravity environment will be available almost indefinitely. Current plans call for three or four shuttle trips to the ISS each year for the purpose of transporting new experiments up and returning samples and results from completed work.[2] Accordingly, NASA anticipates that research efforts will be organized into approximately 100-day groupings, known as "increments," the time between shuttle flights. Assuming adequate storage space for multiple samples, many more investigations can be conducted in an increment than have been conducted on past shuttle flights with science missions. However, since physical items, such as samples or film from cameras, will not be returned to Earth until the shuttle trip at the end of each increment, alternative methods for communicating results to researchers in a timely fashion—such as real-time computer links to data acquisition software on orbit or ground-based control of experiments—will become critical.

The second difference between science on the shuttles and on the ISS is the reduction in crew involvement in experiments. The total number of crew members will not be much different, but the amount of research will have greatly expanded. While the astronaut pool from which crew members will be selected includes people with a

---

[1] EXPRESS racks are skeleton structures that can provide basic resources like power to a variety of modular experiments. Eight separate units, called middeck locker equivalents (MLEs), fit into each EXPRESS rack.

[2] More shuttle trips may occur for resupply and construction purposes.

variety of scientific backgrounds (biologists, physicists, engineers, etc.), it is unlikely that the crew expertise in a given increment will match the large variety of experiments taking place. In addition, through 2005, the primary occupation of all crew members will be ISS assembly. In this situation, any technical innovations that permit the automation of routine tasks or the robotic manipulation of experiments will greatly increase efficiency. The task group also notes that another way to maximize the scientific output of the ISS-based research would be to allow investigators to participate directly in experiments. Scientists could spend time on the ISS as short-term residents or travel with shuttle crews during routine supply and transfer missions. This approach might not be immediately realistic owing to the stringent demands on crew time and expertise during ISS construction. However, as the assembly phase approaches completion, the demands on personnel time should become more flexible, allowing greater crew involvement in research projects and the possibility of having nonastronaut scientists aboard the ISS.

PROTEIN CRYSTAL GROWTH

Since the beginning of the NASA protein crystal growth program in 1985, a variety of equipment has been used to grow and observe crystals in the microgravity environment. Useful and innovative hardware development on systems for the ISS continues today. A complete description of the equipment that is or will soon be available is provided in Appendix A. Options for investigators range from liquid-nitrogen-cooled dewars capable of holding large numbers of samples but providing minimal environmental control or observation to refrigerated trays aligned with Michleson-Morley phase-shift interferometers. The task group was particularly impressed with the prototype of the X-ray Crystallography Facility (XCF), which can grow and cryopreserve a reasonable number of samples as well as provide important monitoring capabilities, such as video feedback and X-ray diffraction data. While this approach of monitored growth appeared to the task group very promising, it is important to recognize that in some situations large numbers of samples might be an effective alternative to a few carefully chosen and observed samples.

The Hardware Development Process

It is clear that NASA's microgravity crystallization program and its associated crystallography hardware are not yet mature. However, it is important that researchers interested in exploiting the microgravity environment on the ISS have access to hardware that is state of the art and the most efficient available. To achieve that goal, collaboration and communication between the various laboratories involved in hardware development should be established. Many of the key pieces of hardware so far have been innovated by external investigators, so it is not necessary to have centralized hardware development within NASA. Multiple developers will encourage variety and creativity, while preventing NASA from getting locked in to a single hardware approach. However, the efforts of hardware developers must be coordinated and communication between them improved to ensure that different programs are not producing instruments with duplicative capabilities and that technological advances are quickly shared and integrated into all equipment where appropriate. In addition, since the modular structure of ISS racks will permit instruments from multiple hardware designers to coexist, it is important that the systems be compatible to allow experimenters to take advantage of the full variety of equipment (e.g., samples could be grown in one type of hardware yet monitored by another developer's system). Finally, the most vital step in hardware development is for the research community at large to have input into the instrumentation development process, as cutting-edge science problems can drive the development of innovative new technologies. An example of the critical nature of the collaboration between science and engineering for effective use of large facilities can be seen in how synchrotron sources and the instrumentation for beam lines have evolved most successfully when bureaucratic structures are not allowed to divorce scientific goals from the work on the technology needed to explore those goals.

The most important characteristic of NASA-sponsored hardware development should be flexibility. If a variety of equipment types are available, investigators, with the help of hardware developers and NASA staff, can match their experiments to the instruments best suited to their needs and goals. A modular approach should be emphasized so that individual systems can be upgraded as technology advances. Finally, NASA needs to be prepared to abandon completed hardware or hardware under development if it becomes clear that better systems or new technologies are available. The evaluation of such hardware should be driven by scientific criteria—"better" hardware is that which is more effective in growing crystals and answering key scientific questions.

Recommendation: The efforts of external hardware developers should be coordinated to ensure that instruments are compatible, to prevent duplication of efforts, to ensure that technical innovations are shared, and to facilitate input from the scientific community in defining the goals and capabilities of protein crystal growth equipment for the ISS. NASA must also be prepared to discontinue development projects that do not use cutting-edge technologies or that are out of tune with the most current scientific goals.

A significant factor affecting equipment development is the instability in the budget for the ISS. Several hardware design and engineering projects have been put on hold as a result of fiscal uncertainties or shortages. If this situation continues, the greatest long-term negative consequence will be the reduction or elimination of a high-quality user community for the NASA protein crystal growth program. This will occur in part because continual uncertainty is demoralizing and discouraging; researchers will avoid a program that cannot offer reliably scheduled research or funding opportunities. In addition, if money is repeatedly siphoned off from the hardware development work, the quality of the equipment on the ISS will be significantly below that of the cutting-edge hardware available on the ground, and researchers will not be interested in using the outdated equipment or willing to entrust precious samples to it. If support originally designated for scientific instrumentation or research becomes a convenient contingency fund for ISS construction, NASA will send a clear message that science on the ISS has a low priority and will alienate the research community even further.

Key Characteristics of Protein Crystal Growth Hardware on the ISS

Designing and performing an experiment for the ISS will be an arduous task for researchers, especially those unfamiliar with the constraints imposed when working in a microgravity environment. The equipment developed by and for NASA should aim to provide a high level of control over samples, equipment, and procedures. On the ISS, crew time will be limited, and the human access to samples and the feedback to the investigators via shuttle trips will be infrequent (once every 3 or 4 months). In this environment, two related hardware characteristics will be essential: automation and ground-based control. Minimizing the number of crew actions required will allow flexibility in scheduling and increase the options open to investigators. In addition, if the principal investigators are able to make decisions about experimental parameters (temperature, time for growth, which samples to preserve or examine) and then are able to adjust experiments in real time from Earth (as they would in their own laboratories), the research produced in each experiment will be of higher quality, and the NASA program will become more attractive. Therefore, hardware development efforts should emphasize the importance of automation, monitoring, real-time feedback, telemanagement, and sample recovery (via mounting and freezing). NASA-sponsored efforts in these areas could be accelerated by increased interactions with engineers developing instrumentation for synchrotron beam lines. This community is interested in very similar issues, including remote control of diffraction experiments, better measures of crystal quality, automation of crystal selection, freezing and mounting, and the rapid characterization of many crystals via routine experiments. Sharing experiences and technologies would benefit both communities.

Effective analysis, preservation, and reentry of promising crystal samples are especially necessary given the key role synchrotrons are playing in protein structure determination. Hardware designers should be sensitive to how synchrotrons define the needs of crystallographers; standard crystal sizes accommodated in ISS equipment should reflect advances in the brillance of synchrotron sources, and techniques for environmental control (pH, temperature, vibrations, etc.), freezing methods, and sample reentry should be fine-tuned to maximize crystal quality. The production and safe return of high-quality crystals from the ISS are crucial if the NASA program is to attract researchers interested in important and challenging biological problems.

While most successful microgravity experiments will probably conclude with diffraction studies at a synchrotron, it is not NASA's responsibility to arrange or guarantee this step. Building a synchrotron beam line is expensive and would not be the most efficient use of NASA's scarce resources. In addition, there are strong user programs at the many beam lines throughout the United States, where user opportunities are allocated by peer review processes. Assuming that NASA's peer review process is selecting the most scientifically rigorous and interesting projects, successful crystallization should enable researchers to compete effectively for the necessary beam time, and success in this extra layer of peer review should further validate the NASA program within the scientific community. Therefore, NASA should not contribute to the construction of a new synchrotron beam line and should not apply for a block grant of time to be reserved for NASA investigators.

The X-ray Crystallography Facility

The X-ray Crystallography Facility (XCF) is a multipurpose facility designed to provide and coordinate all elements of protein crystal growth experiments on the ISS: sample growth, monitoring, mounting, freezing, and X-ray diffraction. A module for the growth phase is designed to house vapor diffusion experiments. The visualization unit uses magnified still photographs of samples that have completed growth to determine whether the resulting crystals are worth preserving. The Crystal Preparation Prime Item (CPPI) is a robotic system that mounts the crystals on hair loops for cryopreservation or on hair loops inside a capillary, unfrozen. Finally, the X-ray diffraction module employs a low-power (24 W) X-ray source and has a maximum resolution of 1.1 Å. The various modules are controlled remotely from the ground; crew time is required to move samples from unit to unit. The various components of XCF are described more fully in Appendix A. The growth phase of research occurs in modular units located in an EXPRESS rack reserved for this purpose, while the CPPI and the X-ray diffraction instrumentation are located in a separate, specially designed rack (see Figure 2.1).

The task group was impressed by the XCF, by the robotics, the remote control, and the range of experimental capabilities provided. The automation is vital on the ISS, where crew time for scientific experiments will be limited and crew expertise may not match up with the experimental tasks. In addition, the ability to observe results in real time and choose the next experimental steps accordingly is attractive, given the length of time between shuttle trips and the limited opportunities for reflight of an experiment. The X-ray diffraction module provides valuable information about whether a given crystal will diffract—this real-time feedback is key to making decisions about the success or failure of a particular crystallization experiment and will help allocate scarce freezer resources by ensuring that the most promising crystals are preserved and returned to Earth. Any diffraction data gained from the ISS X-ray system will be a bonus. The task group offers one specific piece of technical advice: a step in which the sample is swept through a cryopreservative before freezing should be added to the CPPI module; this addition should greatly increase the success rate for freezing.

Finding: Automation, monitoring, real-time feedback, telemanagement, and sample recovery (via mounting and freezing) will be vital for successful protein crystal growth experiments on the ISS. The XCF, through its use of robotics and a variety of experimental and observational capabilities, provides many of the tools researchers need to take full advantage of the microgravity environment.

The task group was impressed by the ground-based control capabilities of the XCF. Current plans call for the technical equipment that performs remote experimental management to be situated at the University of Alabama at Birmingham, rather than at one of the NASA centers. This location is appropriate in the near term, as the expertise of the engineers who developed the system will be invaluable for troubleshooting when the XCF is installed on the ISS, as well as for maintenance and possible future improvements once the system is up and running. It is also important that NASA personnel be involved in these activities. Being allowed to share in the developers' knowledge base about XCF will assist NASA staff in coordinating the XCF instruments with NASA's various other crystal growth hardware and in training principal investigators from a variety of institutions to use the ground-based control systems.

The XCF is typical of several hardware development projects for NASA in that the technologies it employs can be applied to ground-based research capabilities as well as to those based in space. For XCF, the application of robotics to the cryofreezing process and the low-power, compact X-ray source both have the potential to be useful in laboratories on Earth. Currently, however, the scientific community is mostly unaware of the quality of

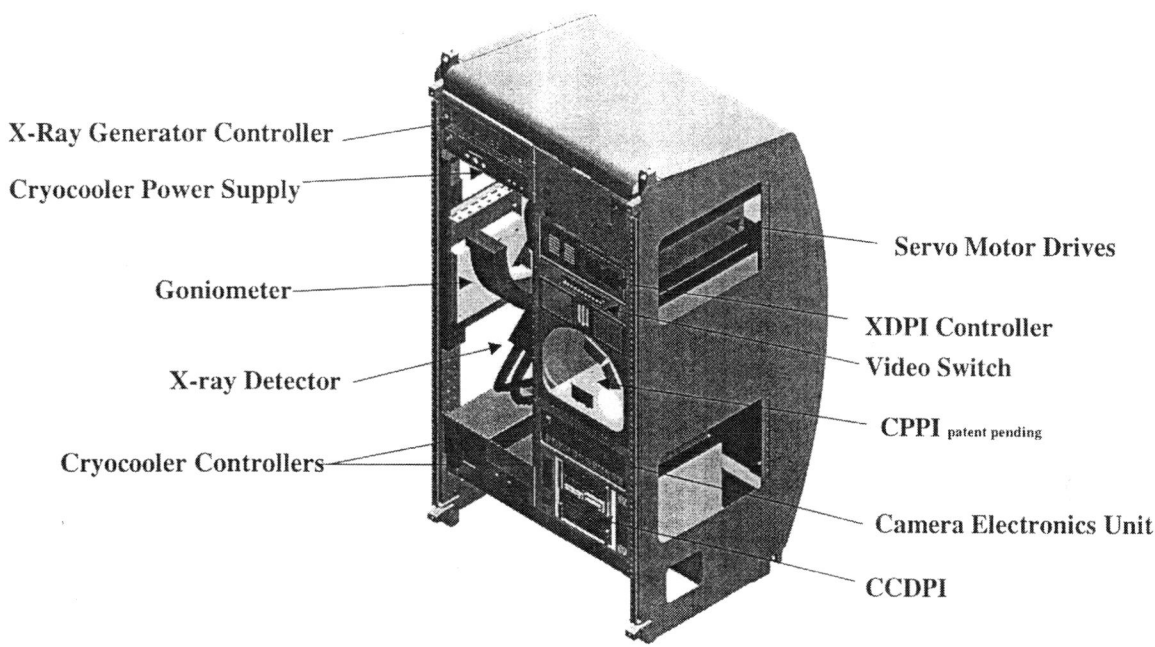

FIGURE 2.1 Schematic diagram of one of the two racks of the X-ray Crystallography Facility (XCF). This rack contains the hardware for mounting and freezing crystal samples and for performing X-ray diffraction studies; the other rack (not shown) contains hardware for growing crystals and observing the growth process. The XCF is in development at the Center for Macromolecular Crystallography at the University of Alabama at Birmingham.

automation displayed in the prototype of the CPPI and of the combined capabilities of the X-ray optics and the low-power source that together yield a high-intensity X-ray beam. While two companies did exhibit this technology at the 1999 Congress of the International Union of Crystallography, the scientific progress made during development of this new product was not widely publicized. While commercial entities may need to be reticient about revealing information about products in development to protect their proprietary work, the task group strongly recommends that full information about technologies and equipment to be used on the ISS be made available to all researchers. This recommendation is based on three factors. The first is that only if the capabilities of the hardware are fully understood will experiments be designed to take advantage of the equipment and the microgravity environment. The second is that well-publicized information about unique or cutting-edge hardware available only through the NASA program will attract a new and broader set of researchers to apply for NASA grants. The third is that scientists will not be willing to entrust precious samples having biologically important implications to equipment unless they are completely aware of both the risks and the potential benefits. The primary goal of the NASA protein crystal growth program should be to serve the research community, not commercial entities. Indeed, NASA's commercial program will be best served by applying the instrumentation to highly visible frontline academic investigations.

CELL SCIENCE

A variety of instruments are being developed to support cell science research on the ISS. Many of them are based on earlier generations of equipment flown on space shuttles or on Mir, but some new approaches are also being investigated. The hardware for cell and tissue culturing falls into three main categories: basic incubators, perfused stationary culture systems, and rotating wall vessels. There is also a variety of supporting equipment, including refrigeration, monitoring, and analytical instruments. A full list and description of various pieces of hardware relevant to cell biology work on the ISS is provided in Appendix A. The task group focused on the most advanced versions of each type of equipment. Its comments in this chapter on the culturing equipment and the support systems take into account the scientific issues facing the research community and the specialized logistical requirements imposed when experiments are performed so far away from the investigators.

Cell and Tissue Culture Hardware

The hardware for cell and tissue culturing falls into three main categories: basic incubators, perfused stationary culture systems, and rotating wall vessels. Currently, NASA intends to use these instruments to support two types of experiments. First, experiments using the perfused rotating bioreactor and, later, a perfused passive system will investigate tissue morphogenesis for the production of normal and neoplastic tissue for research and biomaterial production. Second, the basic incubators and perfused stationary culture systems will be used to understand the cellular response to the changes in force resulting from decreased gravity and to establish microgravity as a tool for understanding fundamental cellular processes. The three instruments described below are all in the late stages of development and scheduled to be deployed between late 2000 and early 2002.

- *Biotechnology Temperature Controller (BTC).* This unit is designed to provide refrigeration on orbit as well as to allow preserving and incubating of multiple cell cultures simultaneously. The cell culture bags are transparent to allow visualization of the samples by light microscopy. While the BTC does not have capability for automated medium exchange, the cultures can be fed using special needleless "penetration" connectors on the bags that provide for multiple aseptic connections. The BTC can accommodate 120 7-ml cultures, or fewer larger samples, within one middeck locker equivalent (MLE).
- *Cell Culture Unit (CCU).* This unit is a modular cassette-style bioreactor that can accommodate multiple cell culture chambers (see Figure 2.2). The CCU provides temperature and pH control and allows for continual feeding and waste medium harvest from perfused stationary cultures (Searby et al., 1998). Mixing occurs via medium recirculation. The CCU also provides automated sample collection and injection and high-quality video microscopy. Individual perfused culture chambers can be replaced on orbit. Specimens are loaded in chambers on the ground; inoculation and subculture can occur in space. Bubbles must be manually prevented from accumulating in the chambers; it would be better if an automated system could be developed to handle this task. The CCU can accommodate 8 large (30 ml) to 24 small (3 ml) samples and the associated support and observation equipment within 2.5 MLEs. This piece of hardware is under development by Payload Systems, Inc., in conjunction with the Massachusetts Institute of Technology for the Life Sciences Division of NASA. The cellular biotechnology program within the Microgravity Sciences Division is funding early development work on a Perfused Stationary Culture System, which is expected to be a small-volume (5 to 50 ml), multivessel system for on-orbit cell culture and tissue engineering investigations. This system is in the early stages of development, has a role similar to that of the CCU, and may not be developed if CCU development is successful.
- *Rotating-Wall Perfused System (RWPS):* This unit houses a single 125-ml rotating wall perfused vessel in a controlled environment along with associated equipment for medium infusion/perfusion, temperature control, gas exchange, and independent wall rotation control (see Figure 2.3 for a schematic diagram of the RWPS's precursor). Unlike ground-based, rotating-wall bioreactors, in which laminar flow is set up to randomize the force vectors and to minimize the shear stress, space-based vessels have rotating walls in order to produce Couette flow, which augments mass transport. Observation and video recording are possible through a large window in the front of the unit. The RWPS can be inoculated on the ground just before launch or on orbit, but once it has been powered and the experiment initiated, it remains powered throughout the increment until landing. Cell and media samples can be removed on orbit through sample ports located on the side and front panels.

Overall, the NASA-funded cell science work to date has emphasized the use of bioreactors to support three-dimensional tissue growth. While the development of rotating-wall vessels has had, and should continue to have, a significant impact on cell and tissue culturing methodology on the ground (see Box 2.1), the task group has a

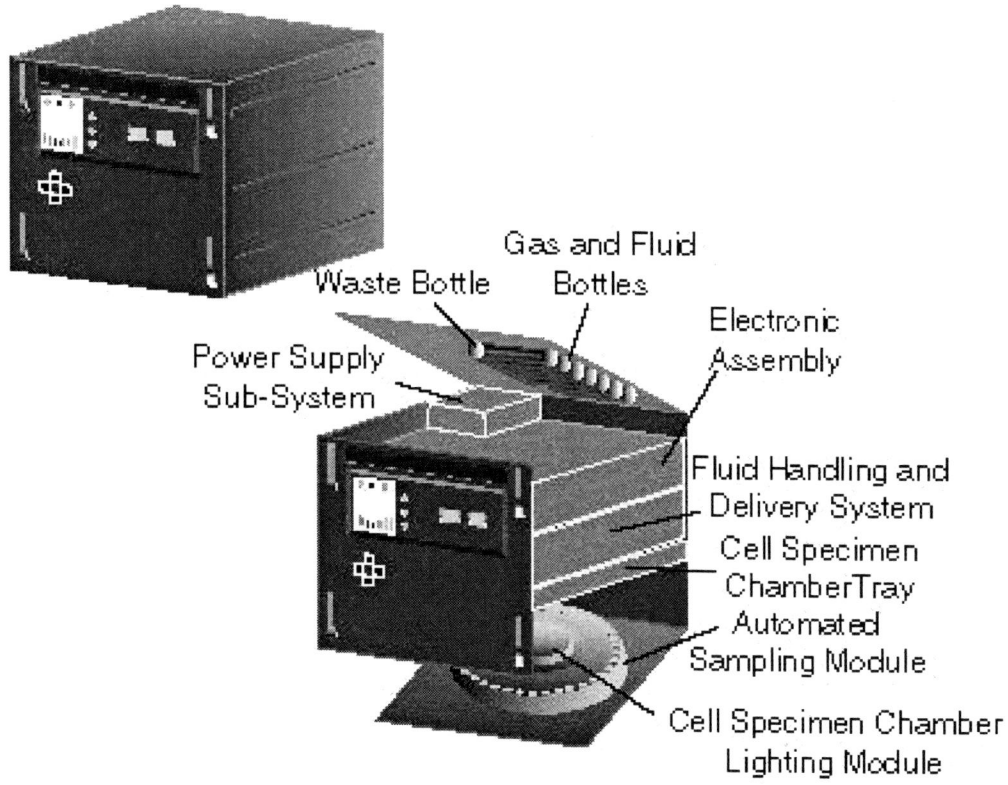

FIGURE 2.2 Schematic of the Cell Culture Unit under development by Payload Systems, Inc. in conjunction with the Massachusetts Institute of Technology for the Life Sciences Division of NASA. Illustration from the NASA Web site, <http://quest.arc.nasa.gov/neuron/photos/images/CCUhabitat.gif>.

variety of concerns about the effectiveness and appropriateness of this approach for research in the microgravity environment. One is that the large volume required for the bioreactor and supporting equipment will yield relatively small amounts of data per unit volume. Other hardware scheduled for use on the ISS (such as the BTC and the CCU) contain smaller specimen units and therefore can serve multiple investigators and house sufficient replicates to permit a complete experiment series in one increment; the potential downside of at least the BTC is the lesser degree of control over environmental conditions. Other concerns about the RWPS relate to the difficulty of accessing the vessel on orbit: sampling is not an easy process and requires a significant amount of equipment manipulation and crew training. Also, bubbles tend to form in the RWPS but cannot be removed without disassembling the entire reactor. A hydrodynamic focusing bioreactor (HFB) is currently being developed at NASA; in this system the shape and rotation of a rotating wall vessel have been redesigned to focus bubbles at one end of the system for easy removal. It may be possible to adapt this approach to allow focusing of cell clusters and, thereby, easy sampling of cell or tissue aggregates of varying sizes from the HFB. If this can be done, the bioreactors could be used as a cell/tissue source for other hardware—e.g., the bioreactor would generate large aggregates that could then be aliquotted and tested in the smaller units for response to a manipulable parameter. Such an approach could generate more data in the same volume based on tissue constructs initiated in a controlled bioreactor. This approach would facilitate the kinetic analysis of cell growth and differentiation processes and would greatly increase the amount of data that could be obtained from a single RWPS reactor experiment. However, the HFB is still in early development, and the current RWPS setup is not appropriately configured for a sample generation role. Furthermore, other ground-based methods for generating three-dimensional tissue constructs, such as the use of scaffolding constructed from biomaterials or micropatterned substrates, may prove to be

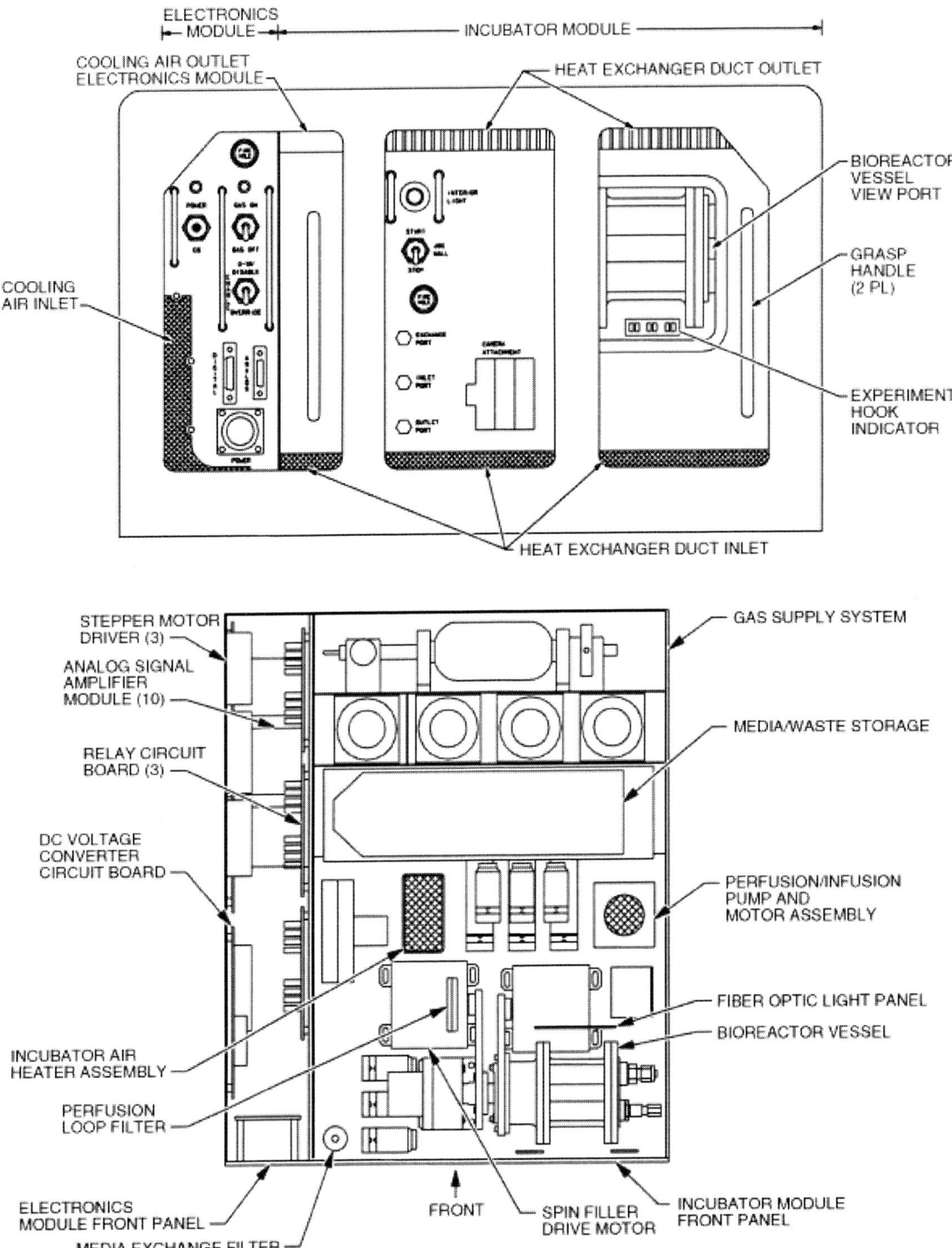

FIGURE 2.3 Schematic diagram of the front view (upper panel) and the top view (lower panel) of the Engineering Development Unit (EDU), the precursor of the Rotating-Wall Perfused System. The EDU has housed rotating-wall vessel experiments on the space shuttle and on Mir. Illustrations from the NASA Web site, <http://spaceflight.nasa.gov/shuttle/archives/sts-70/orbit/payloads/bio/bds/bds201.html> and <http://spaceflight.nasa.gov/shuttle/archives/sts-70/orbit/payloads/bio/bds/bds101.html>.

BOX 2.1 Rotating-Wall Vessels on the Ground

Rotating-wall vessel systems are an important tool for ground-based investigators exploring cellular and tissue responses to low-stress growth environments and simulated microgravity. These systems have been used primarily to produce three-dimensional self-assembling aggregates that retain some of the cell-cell interactions present in tissues. The most important advantage of these bioreactors is the ease with which investigators can form, sample, and feed the aggregates. Several versions of this equipment are available commercially, including a high-aspect rotating vessel (HARV) and a slow-turning lateral vessel (STLV). Scientists at a number of institutions have used the bioreactors to study a wide variety of culture systems, leading to several advances (e.g., propagation of parasites such as cyclospora or those associated with Lyme disease and studies on impaired locomotion of lymphocytes in space). This technology is now mature and new applications are being investigated in a collaboration between NASA and the National Institutes of Health (NIH). Investigators have compiled a large list of tissues that have been propagated in the bioreactor, including cancer cells, cartilage, liver, kidney, lymphoid tissue, thyroid, skin, pancreatic islet cells, neuroendocrine cells, hematopoietic cells, and intestinal epithelium, as well as tissues from the Bowhead whale and microorganisms.[1]

While the rotating-wall vessel systems have been important tools for generating aggregates in cell culture for three-dimensional tissue constructs, the rotating wall vessel is limited in many respects. First, even when co-cultures are used, the tissue synthesized is apt to lack many of the minor cells and elements formed within the intact organism. Second, in cell cultures, the cells that die are not generally removed, creating some artifacts. Third, tissues grown in bioreactors are not subject to the environmental signals that they might sense in situ (growth factors, vascular changes, neuromuscular changes), yet these signals are apt to change in the microgravity environment. In addition to these systemic and environmental drawbacks, the rotating-wall vessel has technical limitations. The limited oxygen transfer capabilities make bioreactors inappropriate for systems with high oxygen demand. Also, it has not yet been determined if rotating-wall vessel bioreactors can provide an appropriate environment for tissues such as osteoblasts that only grow properly when the distances between the cells are maintained. Analyses of which cells propagate better than others and why they do might allow the selection of more appropriate cells for development in the bioreactor and increase the effectiveness of the bioreactor's use for ground-based study. Through a joint program with NIH, NASA has made the bioreactor technology and the expertise of NASA scientists experienced with the hardware available to researchers at NIH, the Food and Drug Administration, and other government laboratories. While this program has expanded the use and appreciation of rotating-wall vessel technologies, a broader outreach program that included scientists from universities and other research institutions might increase the communities' familiarity with the technology and improve understanding of which systems benefit most from the bioreactor's low-shear, low-turbulence environment.

---

[1] Information about the array of projects ongoing under the NIH/NASA interagency agreement can be found in the annual report of the NASA/NIH Center for Three Dimensional Tissue Culture, available online at <http://peer1.idi.usra.edu/peer_review/taskbook/micro/mg99/agreement.html>.

---

more effective sources of samples for multiple-chamber hardware such as the CCU and the BTC. In addition, the amount of data produced by these systems in a given period of time and amount of volume on the ISS will be significantly greater than would be produced by a bioreactor system.

Recommendation: Given the current status of equipment in development, finite fiscal resources at NASA, and the limited amount of volume on the ISS, the task group recommends that future research on the ISS should deemphasize the use of rotating-wall vessel bioreactors, which are already established, and continue to encourage the development of new technologies such as miniaturized culture systems and compact analytical devices.

Another factor to be considered in selecting hardware for cell science research on the ISS is the equipment's ability to contribute to efforts to distinguish between the direct impacts of microgravity, where the low level of gravity alters cell behavior, and indirect effects, where space changes the local culture environment (e.g., variations in gas exchange or nutrient and mass transport rates), which in turn affects the cells. If investigations on the ISS are to tackle this question, the instrumentation should be able to manage and monitor culture conditions, such as cell feeding, environmental stresses, and convective flows, as well as facilitate the comparison of samples with those obtained from a variety of ground- and space-based experimental control groups. For example, methodologies such as micromanipulation could allow researchers to apply controlled mechanical stresses to cells in a microgravity environment to clearly discern whether all mechanosensing capabilities are lost in spaceflight conditions or if the absence of gravity per se has a specific effect on cell behavior.

While the task group believes that of the present instruments, the CCU and the BTC are the most practical tools for research on the ISS, what sort of instrumentation will be most effective for cell and tissue growth in microgravity has yet to be determined. It is important that the relative merits of various pieces of instrumentation be carefully evaluated and that NASA maintain the administrative and engineering flexibility it needs to adopt the most effective systems employing the most advanced technologies and to discontinue hardware development projects that are not attuned to cutting-edge scientific needs of the cell science community. The long-term nature of design and construction of equipment for ISS may be a limiting factor, because the issues and the technology for ground-based research are changing so rapidly that the priorities and hardware selected today may be out of date by the time the cell culture facility is fully operational. Flexibility will be essential, as new culture units, cooling methodologies, and sensors are developed. A modular approach to hardware design must be used to allow for the innovations in technology and the scientific breakthroughs that are likely to occur in the five or more years before the installation of a specialized biotechnology facility on the ISS. Many of the hardware development process issues discussed in the section on protein crystal growth also apply to the development of hardware for cell science. Within NASA's cell science program, close interaction is needed between researchers and in-house operational personnel responsible for developing and constructing hardware to ensure maximum flexibility and responsiveness to evolving scientific goals. Recently, the flight hardware engineering part of the cell science work was placed in an administrative unit separate from that housing the biological research work. Direct communication between these two groups is necessary to construct a facility that can handle relevant, cutting-edge research problems (NRC, 1998). The trend toward separating personnel responsible for the science mission from personnel providing engineering support is, on balance, deleterious to the program. The task group discourages such divisions; instead, NASA should emphasize that the main goal of engineering support is to surmount the challenges faced by space-based science and optimize the research environment on the ISS.

Experiment Management

Cellular systems are very sensitive to environmental perturbations. A continuous power supply to maintain appropriate and stable environments during experiments and for sample storage and transport is essential to ensure valid results.[3] The possibility of brownouts during construction of the ISS and the devastating effects that limitations of power would have on cell culture experiments must count heavily in the selection and design of experiments that will be deployed while the ISS is still being assembled. Because the main activity in this phase will be building on existing structures and attaching new modules, there will be severe vibrations throughout the ISS and significant depletion of an already limited power supply. In addition, the amount of power available for research will fluctuate, increasing every time more solar panels are added and decreasing as new modules are attached. The uncertainty surrounding the power supply (and the variations in the total power available for research) suggests that the shuttle itself might provide a more reliable and predictable base for cell science experiments until the ISS is fully operational. However, volume on the shuttle will be devoted mainly to transporting ISS construction materials during assembly, so research activities will occur only infrequently and in the limited volume available in the shuttle's middeck lockers. Experiments on the shuttle would also be limited with respect to time, and short-term tests of research equipment being developed for the ISS may not supply enough information. Flight of cell science hardware aboard Mir provided crucial lessons about long-term experi-

---

[3]This requirement is relatively unusual among the wide range of research projects being planned for the ISS.

mentation and hardware reliability and limitations. If it would be possible to perform some experiments on the station while ISS hardware is still in development, potential problems with equipment and experiments could be identified and resolved at an early stage. To gather this information effectively, it would be very useful to schedule periods during the construction phase of the ISS where stable power levels are guaranteed.

Even when the ISS is fully operational, management of the culture environment will still be a key issue. NASA is currently developing the Experiment Control System module to provide the interfaces required for communication and control of experimental equipment, execution of the investigators' experimental protocols, and the recording and archiving of experiment and equipment performance data. Work is also continuing on an on-demand control system to manage resources (such as power and gas delivery) so that equipment can exploit scarce resources efficiently and frugally while maintaining appropriate cell culture environments. Such a system would be installed in a specialized biotechnology facility rack dedicated at least in part to cell science work and would juggle resource delivery so that the experiments grouped in the special rack would consume less power than if they were housed in an EXPRESS rack. Also, the system could reroute power within the facility in crisis situations. Since construction on the ISS was just beginning and much of the research equipment was still in development during the course of this study, the task group was unable to quantitatively compare the power availability and the power demand for biotechnology experimentation on the ISS. As plans for volume and resource allotment crystallize in the coming years, careful efforts must be made to coordinate supply and demand for scarce power resources. Even once construction is complete and the Biotechnology Facility has been installed, the modular approach, whereby equipment can be exchanged between increments, will require continued analysis and coordination of the various instruments' power demands.

Another issue that will be problematic particularly during ISS construction but also after the station is complete, is the limited amount of crew time available for research. Crew will often be trained many months before the experiments and personnel are actually flown, and the astronauts' backgrounds cannot always be coordinated with all of the research under way in a given increment. Therefore, a high priority of equipment developers should be to automate routine tasks and to facilitate remote management of the experiments. For automation, the Experiment Control System has the potential to be very useful. For remote management, a variety of factors must be coordinated to allow investigators on the ground to control and adjust their ISS experiments in real time. The development and effective use of robotics for sample procurement and analysis could markedly increase throughput per unit crew time while also increasing data observation by investigators. Since the lead time before space-based experiments and the time before a follow-up experiment can be flown are both long, any approaches that enable investigators to run a complete experiment, including midcourse corrections, within one increment will greatly improve the quality of research performed on the ISS and will make the program significantly more attractive to the scientific community.

Two key supports for ground-based control of ISS research are (1) sensors to enable physiological control of the cell/tissue culture media environment and (2) analytical equipment to provide feedback about the status of cell and tissue samples. NASA on-site contractors are focusing on sensors that will help investigators monitor and adjust their experiments. Development work on sensors for pH and glucose, as well as a pH control system, is quite advanced, while sensors to measure oxygen and carbon dioxide concentration levels are still in the early stages. A commercial off-line blood gas analyzer system will also be available. To ensure that these instruments closely match the scientific tasks of remote experiment monitoring and management, there must be frequent interaction and feedback between NASA scientists and the on-site contractors.

It will be advantageous to have analytical instrumentation on the ISS to characterize the cells that will be subjected to experimental protocols under microgravity conditions. For example, since microgravity affects the cell cycle, a flow cytometer could permit evaluation of cell cycle markers or apoptosis markers or sorting of cells at specific stages. It is recognized that the conventional technologies will have to be modified for such an analytical instrument to operate in the absence of gravity. Since visualization of the cells will be important, in addition to the light microscope that is planned, confocal microscopy (perhaps digital, instead of laser-based) and phase-contrast microscopy would be beneficial, particularly for the analysis of the three-dimensional constructs produced in tissue engineering experiments. (Confocal microscopy might also be useful for protein crystal growth experiments.) Recent innovations such as the "lab-on-a-chip," which utilizes microchip and DNA technologies to analyze biological samples automatically, would enable the miniaturization of facilities, as would devices for single-cell transfer. Gas chromatograph-mass spectrometry will also be important for the comprehensive characterization of natural products.

Because the new generation of analytical instrumentation is very user friendly, requires small samples, and provides rapid output, the potential is very good for significant analytical capacity on the ISS. Recent advances in miniaturization and automation (Henry, 1999) could allow significant enhancements of ISS analytical capabilities with minimal weight and volume penalties. Some Department of Energy facilities, such as the Sandia and Oak Ridge National Laboratories, are developing relevant miniaturized analytical technologies; NASA would benefit from communication with the personnel in those programs. Other relevant technological advances include DARPA-sponsored work[4] on microfabricated substrates and microfluidic systems (biochips) for cell biology. When miniaturized systems are used, many replicas can be analyzed in parallel, more variables can be explored, less weight is required, and many more experiments can be contained within a single unit. In order to efficiently produce the data needed for cell science publications, the equipment on the ISS should enable researchers to perform multiple studies that are repeated on many different occasions and are focused on distinct, but related, aspects of a single scientific question addressed at the single-cell level.

During selection and development of the analytical systems for the ISS, NASA must recognize the vital role of real-time feedback and acknowledge the limited volume available for sample and information storage. The task group was particularly concerned to note that current plans call for both film and digital still cameras and only a general-purpose video camera to be available on the ISS. Since archiving and transmitting data electronically are easier and less expensive, the task group recommends that digital cameras be the primary technology for both video and static microphotography of cell science on the ISS. This approach, when coupled with use of automated analytical techniques (e.g., lab-on-a-chip), will facilitate archiving and allow immediate transmission of information and results to investigators on the ground. Real-time access to data from the sensors and analytical equipment on the ISS is necessary to enable ground-based control of experiments and to provide scientists with real-time records of the progress of experiments. In the construction phases of the ISS, refrigeration and freezer capability and transport space will be limited, so it will be necessary to analyze the samples immediately before and after the experiments. Even when the ISS is complete, preservation, or reentry forces, can change key characteristics of the samples, so on-orbit analytical capabilities will continue to be essential. If digitized data from on-orbit analyses were available to investigators in real time, they would be less dependent on shuttle trips for results, could select the most important samples for the scarce storage space, and study the changes wrought in samples by freezing and reentry. The value of real-time data should not be underestimated, but neither should digital data from on-orbit sensors be considered a comprehensive substitute for the information that can be gathered on samples when they are returned to Earth and undergo thorough investigation with the wide array of instruments available in ground-based laboratories.

Finding: The limited amount of crew time available for research-related work and the infrequency with which investigators will have access to their samples via shuttle trips mean that automation of routine tasks, ground-based control of experiments, on-orbit analytical capabilities, and real-time transmission of digital data are vital for conducting effective cell science research on the ISS.

It is important to recognize that, while equipment is being designed now, the primary use of the cell science facilities on the ISS will begin 2 to 5 years into the future, and researchers will require access to the most up-to-date analytical instrumentation for their space-based experiments. The focus on quantitative measures of cell and tissue behavior in microgravity will require new forms of genomic, proteomic, and metabolic analysis. Relevant instruments could include fluorescence microscopes using both real-time and time-lapse imaging at up to 100×

---

[4]The DARPA MicroFlumes program is described on the Web at <http://www.darpa.mil/MTO/mFlumes/>.

magnification in order to take advantage of techniques that analyze dynamic changes in cell structure and function. Microinjection, micromanipulation, microfluorimetry, use of optical tweezers, and magnetic manipulation are other potential methods to consider as ways to address many of the mechanistic questions relating to gravisensing in cells.

Storage, Transport, and Throughput of Samples

In the initial stages of ISS construction, the generation and retrieval of samples will be limited by the availibity of refrigeration volume. In the absence of low-temperature refrigeration, initial studies will require fixation, and perhaps storage, at ambient temperatures. This approach will require additional ground controls. During flight, multiple experiments will have to be run concurrently, but even comparisons between these systems on orbit will not be able to isolate the effects of launch on the cell and tissue cultures. This limitation will force investigators to have analyses performed immediately after the cultures arrive on the ISS, and then again at predetermined intervals. Reentry will cause other problems, making facilities for on-station analyses doubly useful. Before analytical capabilities are fully installed on the ISS, preservation of samples will be especially important, so the task group recommends a more detailed investigation of methods to optimize sample preservation and recovery in the construction phases of the ISS. Since the paucity of freezer and storage space will impose serious limitations on the type and number of experiments performed, the task group believes that higher priority should be given to acquiring a freezer or other device (e.g., a dewar of liquid nitrogen) for transporting and storing cryopreserved cells during the early phases of ISS construction. This will allow cultures to be initiated on-orbit, eliminating launch effects and increasing reproducibility and throughput. For example, cryocapabilities will permit scientists to assess the influences of varied gravitational forces on parameters of cell structure and function.

Once the ISS is complete, plans call for a variety of refrigeration capabilities: cryogenic (–183°C), fixed samples (–80°C), reagents (–20°C), and storage systems (+4°C). To maximize throughput, plans call for preparing cultures from frozen batches of cells, so these refrigerators and freezers are needed for storing cells for the inoculation of cultures in the middle of an increment. There is some concern that the amount of storage or transport facilities might not be adequate for a sufficient throughput of samples to support an active and viable research program. To make the most of limited volume, miniaturization of samples (even single-cell studies) might be considered. Effective miniaturization of experimental hardware could substantially increase the number of samples analyzed and, hence, the amount of data generated and the number of primary articles published per year.

Another factor affecting the throughput of cell science experiments at all phases of ISS construction and use is access, the ability to get samples and equipment to and from the station. Cell science work requires late loading (14 to 24 hours before launch) and early unloading (3 to 8 hours after landing) from the shuttles to and from the ISS. Cell science samples must also be transported on the shuttle middeck. This will limit access to the ISS in the early phases, when middeck space is at a premium. Thus, there is an incentive to transport specimens that require tight temperature control separately from large equipment such as the CCU and BTC modules, which could fly in the shuttle bay or even be left on the ISS. In this scenario, specimen loading and removal would occur on-orbit, assuming crew time would be available for sample transfer while the shuttle is docked. One factor that must be considered in transporting equipment in the shuttle bay is that the bay can reach 120°F during launch.

Refrigeration and transport are not the only factors limiting the throughput of cell science research on the ISS. Other factors include crew time required for the experiments, the amount and reliability of the power supply, adequate storage space and appropriate environments for samples and supplies, shuttle flight schedules to and from the ISS, the volume of materials to be transported, and, of course, the size of the budget provided for cell science hardware development and research support. Many of these factors are affected by competition from other activities for resources on the ISS, and some of the problems are not specific to cell science but affect programs throughout NASA.

Current plans call for NASA to support between 5 and 15 cell science flight investigations on the ISS each year, and it has been suggested that this level of activity should result in the publication of between two and five primary articles each year. The task group believes that the impact of the papers is more important than the number of papers. A window of opportunity has been created by the advances in molecular, cellular, and biochemical approaches (e.g., functional genomics and proteomics) that are occurring as the ISS research platform becomes available. The task group recommends that to most efficiently exploit this opportunity, emphasis should be placed on integration of the different approaches and on collaboration between principal investigators and other intra- and extra-NASA investigators. The impact of a given number of experiments could be increased by greater collaboration between investigators interested in different aspects of the same experiment and by more effective management of experiments. For example, a portion of the tissue samples from experiments designed to evaluate the effects of microgravity on three-dimensional tissue formation could be analyzed for changes in gene expression by a collaborating investigator. In this regard, it may make sense to issue NASA Research Announcements (NRAs) requesting proposals for additional analyses that could be piggybacked onto planned flight experiments (these would be described in the NRA). It may not even be necessary to fully fund these piggyback investigations, as the costs associated with flight would be covered under the primary investigator's grant and the main cost to the secondary researcher would be for sample analyses. Of course, care would have to be taken to avoid conflicts between different experimental protocols. Collaborations between the Life Sciences Division and the cell science investigators could also enhance the scientific output of space station research; these opportunities should be advertised on the NASA Web site to encourage cooperative efforts. NRAs for flight experiments should be modified to encourage applicants to establish multiinvestigator collaborative projects in advance. If this is done, part of the proposal evaluation should include assessing the total amount of information that would be obtained from a flight experiment, including that from collaborators.

Recommendation: Mechanisms should be developed to enable collaborative research projects that maximize the amount of data obtained from each cell or tissue sample by executing multiple analyses on each sample.

Even after experiments are complete and have returned to Earth, methods for archiving the results will be an additional challenge. Current plans do not call for the retrieval/reentry of viable samples from the ISS, but preserved samples and the data obtained both on orbit and during post-flight analysis, such as gene sequences, would be valuable resources for the scientific community. It is important that samples and data from all experiments be archived and the information made readily available, particularly for unsuccessful work that does not result in published articles but probably contains lessons about the difficulties of on-orbit research. Issues related to archiving include what kind of storage will be available, who will own data such as gene sequences (especially data obtained in international collaborations), and how access to archived specimens and data will be advertised to the general science community.

OVERALL VOLUME ALLOTMENT FOR BIOTECHNOLOGY RESEARCH ON THE ISS

Currently, plans call for biotechnology research funded by NRAs to occur within one rack on the ISS. This rack would be shared by protein crystal growth and cell science work. In addition, two racks are reserved for the hardware associated with the X-ray Crystallography Facility (XCF). The task group considered this arrangement and the needs of the various research communities and recommends a shift in the allotments. Of the two XCF racks, one is filled with equipment related to the automated cryofreezing of protein crystal samples and the X-ray diffraction of these samples. The other rack is devoted to growth of the crystals and visual monitoring of the results, and is currently reserved for the equipment and users of XCF and the Space Product Development Division. The task group recommends that this latter rack be officially dedicated to the protein crystal growth program of NASA's Microgravity Research Division, where experiments are selected by a centralized peer-review process and where all protein crystal growth hardware can be used. The rack currently scheduled to be shared by cell science and protein crystal growth can then be dedicated entirely to cell science research.

The task group makes this recommendation based on several considerations. A primary issue is the basic incompatibility between the technical needs of cell science and protein crystal growth equipment on the ISS. For the culture growth hardware, the primary concern is environmental control, which is most effectively maintained by connecting the cell science experiments to the ISS cooling loop to take advantage of ISS-wide efficiencies.[5] For the protein crystal growth experiments, minimal vibration is important and can be obtained by installing an active rack isolation system (ARIS), which cushions the rack from ISS motion and the effects of surrounding equipment. ARIS and the cooling loop cannot be installed on the same rack. The cell science experiments require frequent exchanges of samples, which would also negatively affect the vibrational quiet of the protein crystal growth modules. If cell science and protein crystal growth equipment are housed within one rack, one or both of the disciplines will be forced to operate under suboptimal conditions. Another technical consideration is coordination with other related equipment on the ISS. It would be more efficient for the cell science work to be located near associated analytical equipment or storage facilities, while the importance of the monitoring, mounting, and freezing capabilities of the XCF means that the protein crystal growth experiments need to be closely coordinated with this facility.

The task group also carefully considered the needs of the various research communities expected to use the biotechnology facilities on the ISS. For cell science, there was concern that the amount of data and results generated from half a rack would not be substantial enough to maintain interest within the scientific community, whereas a full rack's worth of instrumentation could raise the program to a critical threshold. For protein crystal growth, the research community is still uncertain about the benefits of growing crystals in a microgravity environment, so the guest investigator program is undersubscribed and commercial interest is low. By focusing the protein crystal growth research efforts on biologically challenging problems and by emphasizing hardware capable of monitoring and preserving samples (the XCF, for instance), NASA could direct its resources to validating the program. The current volume commitment of half a rack of general macromolecular research is insufficient to establish the value of the crystal growth program, but a full rack, filled with peer-reviewed experiments that employ all types of available hardware and have access to the capabilities of the XCF, should be adequate to give the program a fair chance of success. If after several years the results from the protein crystal growth work have provided sufficient proof of microgravity's benefits and the academic and commercial demand for facilities on the ISS increases, then high-throughput hardware should be developed and the allotment of space on the ISS reconsidered based not only on the demand for macromolecular crystallography research volume but also on the results to that point from the cell science program. Alternatively, if the work done through the augmented commitment suggested here fails to clearly demonstrate the value of microgravity for work in structural biology, then the protein crystal growth program can justifiably be terminated.

Recommendation: The volume allotment for biotechnology work on the ISS should be redistributed as follows:

- *The mounting, freezing, and diffracting equipment of the X-ray Crystallography Facility (XCF) should occupy one rack (as currently planned).*
- *The cell science work should occupy the entirety of what is currently designated the Biotechnology Facility.*
- *The rack currently assigned to the XCF growth equipment and managed by NASA Space Product Development should be officially dedicated to the peer-reviewed macromolecular research run out of the NASA Microgravity Research Division.*

---

[5]This loop circulates water at low (3 to 6°C) and moderate (16 to 18°C) temperatures; the water flows through heat exchangers attached to individual payloads in order to reject heat generated by the various experimental instruments.

3

Selection and Outreach

SELECTION PROCESS, OUTREACH EFFORTS, AND COMMUNICATION AMONG PROGRAM PARTICIPANTS

NASA research in cell science and protein crystal growth is funded through a collection of 4-year grants. As of 1999, NASA's biotechnology research grants totaled approximately $19 million dollars per year. The roughly 90 principal investigators supported by this program are split evenly between cell science and protein crystal growth; each project receives about $200,000 per year. Applications for these grants are solicited every other year via a NASA Research Announcement (NRA) released by the biotechnology section of NASA's Microgravity Research Division. The focus of the NRAs is developed with advice from the extramural science working groups, and both ground-based and flight projects are considered. NRAs are released to the science community through mailings, the Internet, and announcements in a variety of journals. Proposals are submitted to NASA headquarters, which convenes several peer-review panels to evaluate and rank the applications based on scientific merit. For applications proposing flight experiments or flight hardware development, NASA staff convene a second level of review to focus specifically on issues related to the design of experiments for space. Once approved, ground-based research grants are monitored on an annual basis by NASA's program personnel at the Marshall and Johnson Space Flight Centers; each project submits an annual report and a list of publications, and the investigators jointly attend and present results at a NASA-organized meeting. Flight investigations are monitored and assessed for flight hardware requirements by NASA staff at headquarters and at the space flight centers. Flight experiments are subjected to a Science Concept Review to determine the need for microgravity, the availability of hardware, and resource requirements. NASA staff help flight investigators define protocols, simplify and minimize requirements for crew time, and prepare for flight. Selected experiments are queued and flown based on volume availability and flight specifications and capabilities.

Improving the Dissemination of NRAs and NASA Program Results

In the most recent biotechnology NRA (opportunity announced in December 1997; selections made in November 1998), there were 165 proposals, of which 48 (29%) were funded. While the NRA mechanism is appropriate, it is inadequate to attract the involvement of the best scientists or bioengineers. The task group believes that as the program goes forward, it would benefit from a strengthening of the outreach, selection, and support offered by NASA to ensure that the proposals submitted for consideration are of the highest quality and that everything possible is done to give flight experiments the best chance of success. On the following pages, the task group offers observations on and suggestions for improving the processes associated with selecting and executing research in space; another NRC report (NRC, 1999) has provided a discussion of an administrative institution designed to facilitate research on the ISS.

As NRAs are released, more should be done to disseminate them widely to the communities that might be interested in using NASA biotechnology facilities on the ISS. Protein crystal growth scientists and cell science researchers, but especially the latter, identify themselves with a variety of different professional organizations, publications, and conferences. Currently, the NRAs are sent to NASA's own mailing list, as well as to lists obtained from the Biophysical Society, the Society for In Vitro Biology, and Protein Science. Notices about the NRAs are also posted in *Nature* and *Science* and in publications of the American Society for Cell Biology and the Federation of American Societies for Experimental Biology. There are many other publications through which funding opportunities could be communicated. For protein crystal growth, some examples are the newsletters of the International Union of Crystallography (IUCr) and the American Crystallographic Association. For cell science, possibilities include journals such as *Trends in Cell Biology* and the newsletters of organizations such as the American Institute of Chemical Engineers' Food, Pharmaceutical, and Bioengineering Division, the American Chemical Society divisions of Biochemical Technology and Biological Chemistry, the Biomedical Engineering Society, the American Society for Bone and Mineral Research, and the American Society for Biochemistry and Molecular Biology. Although the NRAs are currently posted on the Internet, NASA maintains so many Web pages that the online NRA is probably useful more as a reference for those who are already aware of the opportunity than as an introduction to the program. Another approach to expanding the pool of potential researchers would be to issue NRAs in collaboration with other federal agencies, such as the National Institutes of Health (NIH), the Biotechnology Program in the Engineering Directorate of the National Science Foundation (NSF), the NSF Biological Sciences and Regulatory Biology Divisions, and the Department of Energy. Cross-cutting projects could also be supported via memos of understanding with these other government agencies.

In addition to broadening the dissemination of NRAs, more could be done to provide sufficient background information for potential investigators who are not familiar with NASA programs. In particular, new investigators interested in flight experiments typically need to be introduced to the special opportunities and constraints of ISS-based research. A lack of familiarity may inhibit some investigators from submitting proposals and may also decrease the probability that paradigm-challenging proposals will be submitted. This difficulty in recruiting new investigators could be alleviated by NASA's hosting biotechnology workshops at relevant, widely attended national and international meetings. The task group recommends that these workshops focus not only on the biotechnology research topics currently supported by NASA but also on the hardware available for experiments on the ISS. Potential new investigators need this information about the specialized systems available and the limitations of space-based research in order to craft proposals that can reasonably be expected to reach the flight definition stage of grant review. More information on hardware opportunities, limitations, and constraints would assist new and experienced investigators in efficient and effective experimental design.

Another way to increase the quality and quantity of proposals submitted in response to NRAs is to more broadly disseminate the results of completed and in-progress NASA-funded work. The most comprehensive presentation of recent results in cell science is at the Investigators Working Group (IWG) annual meeting. Attendance at this meeting has grown rapidly; however, the attendees are already familiar with the NASA program and are composed exclusively of researchers and staff involved in the cell science work funded by NASA's Microgravity Research Division; broadening membership in the IWG to include investigators and NASA staff in the Life Sciences Division would increase cross-fertilization. Also, the task group recommends that NASA consider inviting outside speakers and guests to the IWG meetings to extend the reach of the presentations. Another mechanism to increase potential user awareness would be to encourage NASA-funded researchers to speak at a wider range of conferences. The task group cautions, however, that care must be exercised to ensure that presentations give a balanced portrayal of successes and limitations so as not to raise unrealistic expectations. Incorrect perceptions of the NASA programs can also arise from press releases that target the general public and portray potential future applications of the areas under study in NASA-funded fields as completed or current work. This publicity often leads to misconceptions about NASA's goals (the cell science program does not aim to grow artificial human organs in space) or accomplishments (the protein crystal growth work has not produced a new flu vaccine). NASA is a federally funded agency, so the importance of its work does need to be communicated to Congress and the public, but by allowing the widespread dissemination of vague or even inaccurate descriptions of the programs, NASA is seriously diminishing the credibility of its work within the scientific community.

In the past 2 years, NASA protein crystal growth personnel have attended the American Crystallographic Association Annual Meeting, the IUCr Congress, the International Conference on Crystallization of Biological Macromolecules, the annual meeting of the Federation of Analytical and Spectroscopy Societies, and the Annual Meeting of the American Society for Cell Biology. Other meetings that attract audiences who would be interested in the results of NASA programs include the annual meetings of the Protein Society and the American Society for Biochemistry and Molecular Biology and the Gordon conferences on protein chemistry and structural biology.

In cell science, much of the current ground-based scientific program is of high quality, and early flight-based results are intriguing and promising. However, the larger scientific community is generally unaware of both the quality of this recent work (since much has only recently been submitted or is in press) and the opportunities for future projects. The impression is that the cell science program in NASA's Microgravity Research Division has focused on generating three-dimensional tissue constructs for potential commercialization. From the presentations to the task group and the latest NRA, it is clear that NASA is actually interested in a broader program in cell science and in moving beyond the qualitative phase of observational science in tissue constructs to probe the underlying biological and engineering questions. NASA personnel have attended the annual meeting of the American Society for Cell Biology, the general meeting of the American Society for Microbiology, the Congress on In Vitro Biology, NASA/Juvenile Diabetes Foundation technology workshops, and the Space Technology and Applications international forum. Other conferences at which NASA could expand its audience include the Engineering Foundation Conference on Cell Engineering, the Tissue Engineering Society meeting, the Society for Biomaterials meetings, and meetings of the American Chemical Society's Division of Biochemical Technology and of the American Institute of Chemical Engineers' Food, Pharmaceutical, and Bioengineering Division. Also worth considering for work related to bone and cartilage cells are the American Society for Bone and Mineral Research, the Orthopedic Research Society, the Arthritis Foundation, the American Association for Aging Research, the National Osteoporosis Foundation, and the International Association for Dental Research.

Another important element in increasing the quality of experiment design for NRA-inspired proposals is to more widely disseminate information about past negative results. While information about successful experiments is likely to be published, negative results are often unavailable. The task group recommends that NASA make information about unsuccessful experiences available via the World Wide Web and reference this source of potential difficulties in the NRAs. Making this information more widely available would allow researchers to focus their time and effort on new ideas with a higher probability of success.

Recommendation: NASA should improve its outreach activities in order to involve a broader segment of the scientific community in its biotechnology research program and to increase the number of cutting-edge projects submitted for funding. It needs to disseminate NRAs and program results more widely and to provide more complete background information on failed projects and how to design flight experiments.

Improving the Selection Process

As the pool of applicants expands, the process of evaluating proposals may need to be adjusted. Some concerns about the current process include the 2-year gap between NRA grant submission opportunities.[1] This schedule is likely to inhibit applications directed at the most cutting-edge research issues. In addition, the 2-year gap may discourage the resubmission of proposals that need revisions. Another problem with NASA's current system is that after funding, the delay between project selection and flight manifesting of an experiment may dissuade even a researcher whose first experience has been successful from instigating a follow-up project. Such delays also mean that NASA does not always have the hardware flexibility to respond to changes in the field based on new developments in ground-based research (such as the increased reliance on cryoprotection and freezing of

---

[1] As a point of comparison, the NIH accepts proposals throughout the year, with review groups convening every 4 months. Successful projects are awarded funding within 9 months of the date of submission.

crystals or the use of scaffolding for three-dimensional tissue constructs). Finally, the uncertainties surrounding the NASA budget and the continual schedule changes make people cautious about getting involved in a program that is unable to reliably predict the availability of money or the schedule for access to the ISS. These fiscal uncertainties also affect the quality of NASA's internal programs, because periods with low funding may result in the permanent loss of experienced contract workers to other positions and projects.

Improving Connections to Relevant Communities and Attracting the Best Science

One critical step toward raising the profile of the NASA program and the quality of the grant application pool would be to counter the current perception of recipients of NASA funds as a closed community with a fixed membership. The same names tend to appear repeatedly on NASA lists of grantees, advisory groups, peer review panels, and institutes. For example, the scientists in the virtual NASA Space Biomedical Research Institute seem to be concentrated in only a few institutions. Also, in the protein crystal growth's Guest Investigator Program, access to flight is not centralized and arrangements depend in part on connections with the principal investigators who developed the relevant hardware.

On the whole, external input into NASA's priorities for the biotechnology program seems to be relatively limited. Current advisory mechanisms include the extramural Science Working Groups involved in developing research announcements, panels formed by NASA headquarters to peer-review grant applications, and the Biotechnology Discipline Working Group (DWG).[2] These groups are currently composed of many of the same people who make up the pool of grantees, contributing to the perception that NASA is not really interested in input from outside. By reaching out to a broader slice of the protein crystal growth and cell science communities, NASA would not only increase the quality of the advice it receives but would also be able to educate a new group of people about its programs. One of the difficulties that comes from the lack of a single appropriate forum (conference, association) for a program-wide evaluation of NASA protein crystal growth or cell science work by the community is that sometimes NASA personnel are not in contact with all of the right people in all of the right fields. Expanding the groups represented on NASA advisory panels should help resolve this issue.

According to NASA, the Biotechnology DWG is currently the main mechanism for receiving advice about the strategic direction of the Microgravity Research Division's biotechnology programs. It has been difficult for NASA to attract prominent outside researchers to the DWG. The task group offers two recommendations to address this difficulty. One problem is the name, which masks the group's high-level advisory role. It is recommended that the name be changed, perhaps to "Research Advisory Panel" or "Strategic Planning Committee." The other problem is the mixing of protein crystal growth and cell science within one committee. It is recommended that the DWG be split into two separate groups. This would allow each panel to focus on the issues most relevant to its respective scientific area. The greater number of slots available for each area would also give the panels greater breadth. Care must be taken in selecting new members to ensure that there is not a bias toward those already working with the NASA program; a new name for the DWG might help to attract outsiders of high stature in the macromolecular structure determination community and the cell science community. The revamped groups should be able to provide more useful advice on a variety of issues related to the scope of research announcements, peer review practices, and future programmatic directions.

The importance of recognizing the differences between the two facets of NASA's biotechnology program and allowing each to develop its own identity is also relevant to NASA's outreach efforts. The current practice is to include both protein crystal growth and cell science in the same NRAs, publicity releases, and meeting events, such as the session at the annual meeting in 1999 of the American Society for Cell Biology. The task group believes that outreach efforts would be more effective if separate workshops, presentations, and opportunity notices were used to communicate with the wide array of communities that could be affected by and involved in

---

[2]The program also receives occasional input in the form of reports published by National Research Council committees, such as the standing Committee on Microgravity Research.

the biotechnology work supported by NASA. The composition of the peer-review panels may also need to be adjusted. For example, as the cell science program moves forward with a focus on fundamental research, the task group believes that the representation of bioengineers on the peer-review panels should be increased (from 1991 to 1998, bioengineers made up only about 10% of the cell scientists on the panels).

Recommendation: The separate identities of the protein crystal growth and cell science sections of NASA's biotechnology research program should be emphasized. One key step should be splitting the Discipline Working Group into two strategic advisory committees to reflect the different issues facing each area of research. Prominent scientists not familiar with NASA's programs but aware of the broader issues facing the fields should be recruited to serve on these committees.

One element that may help to broaden the community involved in space-based biotechnology research is the international nature of the ISS. In addition to the facilities described in this report, an array of protein crystal growth and cell science equipment is being developed for the ISS by NASA's international partners, specifically by the European Space Agency and by the National Space Development Agency of Japan.[3] While one country's space agency will not directly fund foreign researchers, bilateral interagency agreements to share hardware are possible, and international announcements of opportunity are planned. Currently, the International Microgravity Strategic Planning Group, which includes representatives from the U.S., Canadian, Japanese, European, Italian, French, and German space agencies, has been meeting two or three times a year to discuss cooperative efforts on hardware development and strategic plans for microgravity research on the ISS. Work on international announcements of opportunity is under way; an announcement specifically about microgravity research on biotechnology is scheduled for 2001.

Coordination: Investigators and Operations Personnel

An important issue for execution of research in the unforgiving environment of space is the potential for conflict between the scientific goals of an experiment and the engineering limitations associated with a space-based platform like the ISS. Many scientists in the biotechnology community believe that in the face of such conflicts, NASA has always opted for ease of operation over support of science. Indeed, the perception is that the NASA culture does not emphasize the importance of communication between scientists and operations personnel, nor does it provide tangible assurances to the research community that the execution of high-quality research in hardware designed to answer the most cutting-edge scientific questions is a NASA priority. The task group believes that to enable important research to be performed effectively on the ISS and to attract the best scientists to the NASA program, this perception must be changed. For example, the task group learned that the computers and operating systems to be used on the ISS for experiment manipulation and data storage have already been determined. While it is understandable that a uniform interface will make astronaut training easier and simplify system engineering and design, the rapidly changing nature of computer technology and the wide variety of systems in use by outside investigators indicate that some flexibility on this point not only would permit researchers to select and tailor computer systems to their specific experiments but also would allow NASA to utilize the most advanced technologies on the ISS.

Biotechnology is a rapidly evolving field, and the key questions in both cell science and protein crystal growth are changing quickly. For biotechnology facilities on the ISS to be effective, a flexible, modular system is required to allow for changing scientific needs and to take advantage of ground-based technological innovations. The task group recognizes that the unique constraints of space-based research, including the need to plan experiments and develop hardware far in advance in order to schedule astronauts' training and resource allocation on the ISS, may

---

[3]For links to descriptions of the European and Japanese biotechnology facilities, see <http://www.science.sp-agency.ca/K3-IMSPG-Facilities.htm>.

limit NASA's ability to respond to changing scientific goals. However, technological breakthroughs often occur because cutting-edge scientific problems require new equipment, so close collaboration between investigators and the operations personnel responsible for fabricating hardware and planning flight manifests will benefit all aspects of the program. NASA personnel or contractors responsible for infrastructure development and operations have no incentive to make scientific goals their priority unless scientists evaluate their performance and their products (hardware, support functions, etc.) based on how well they have advanced the science.

Another factor in attracting the best investigators to the program is the ability to create an environment in which the investigators are confident that their flight experiments will be successfully carried out and that they will be able to observe and modify their experiments on orbit. Two important communication issues need more attention: (1) communication between astronauts and investigators about the analysis of results and experiment manipulation and (2) communication between investigators and decision makers in times of ISS resource adjustment and crises. The first area includes not only preflight training but also telemanagement of experiments and coordination during flight. In its discussions of instrumentation for both protein crystal growth and cell science, the task group emphasizes the value of remote management of experiments by automating routine tasks and developing specialized robotics. The second area is particularly vital for cell science work, where continuous power is needed to maintain environmental control and crew time is needed to perform a variety of tasks that have not been automated. A system that enables efficient communication between the NASA operations staff making resource allotments and the investigators flying experiments is one important way of attracting qualified researchers to the program and maximizing their ability to perform successful experiments. This issue is discussed further in the cell science section in this chapter.

Recommendation: The NASA culture tends to limit communication and coordination between operations personnel and researchers during hardware development; between astronauts and investigators before and during experiment execution; and between decision makers and scientists about the allotment of resources in times of crisis. To attract the best investigators to its biotechnology program, NASA must create an environment geared toward maximizing their ability to perform successful experiments.

PROTEIN CRYSTAL GROWTH

The Guest Investigator Program

In addition to NRAs and hardware development grants, the NASA protein crystal growth program created a guest investigator program to involve external investigators in flight experiments. In this program, when a NASA principal investigator is manifested on a flight, he is authorized to recruit other scientists (guest investigators) interested in providing macromolecular samples to be crystallized in space. No fiscal support is offered, only the opportunity for flight. Interested guest investigators are then screened by a small group consisting of the principal investigator, NASA staff, and one other scientist experienced with flight experiments. Between 1985 and 1998, 67 individuals participated in the program, with a total of 147 guest investigations occurring in that time period. While this program has provided hardware developers with useful information about the capabilities of new equipment, it does have several flaws that limit its effectiveness. As the program is channeled through individual principal investigators, there is no opportunity for centralized review of proposals and guest investigators are limited to the hardware being flown by the principal investigator recruiting them into the program. In addition, without formal grant agreements, there is no mechanism to ensure that appropriate experimental controls are studied in tandem with the space experiments or that all results, both positive and negative, are reported. These constraints suggest that the program's focus is more on testing equipment and less on providing an environment to investigate protein crystal growth in microgravity or solve important structure determination problems.

As discussed in Chapter 1, some of the results that have come out of the guest investigator program (such as the work on EcoRI-DNA and on insulin) indicate that microgravity has the potential to be an important tool for crystallization. Overall, however, the set of data produced in the 13 years of this program has failed to definitively demonstrate that space-based crystallization can be a key step in structure determination research. The program has been undersubscribed, and the level of control allotted to the principal investigators and the relatively limited number of external investigators involved have reinforced the belief of many researchers that the NASA program included only a small, closed community. The task group recommends that the current guest investigator program be phased out. Hardware developers should still be given some flexibility and allowed input into some of the samples that are flown while the equipment is on testing flights, but in general control over which protein crystallization efforts are flown should be centralized at NASA, where a peer-review process should be employed to build a coherent program with clear strategic goals.

Funding Research on Biologically Challenging Problems

The focus of the NASA program should be on demonstrating microgravity's effect on protein crystal growth. To that end, the task group proposes that NASA instigate a high-profile, nationwide series of grants. These would be 3-year grants, in which investigators would receive roughly $200,000 per year, and NASA should award only five each year (for 5 years). (The total amount given out in the 25 grants would be $600,000 per grant for a total of $15 million over 7 years.) These grants would be given to researchers to support simultaneous efforts to get the best possible crystal on the ground and the best possible crystal in space for biologically important macromolecules. (To make the opportunity attractive, NASA would have to guarantee that at least one flight opportunity would occur within the grant period, preferably in the last year.) If in several cases out of the 25 the space-grown crystal turns out to be better and enables the solution of a hot scientific problem, the microgravity crystallization program will have been validated. Also, the scientists who experienced success will be good spokespeople for the program (especially if they had no previous microgravity experience or connection with the NASA program).

The task group believes that this is the best way for NASA to validate the program and to obtain community buy-in. In research on the structure determination of biologically important macromolecules, the crystallization samples take considerable time, money, and effort to produce and hence are very precious to the researchers, who will need some incentive to engage in a risky research environment (space) where their control over the samples is limited. A generous grant would provide such an incentive and lower the risk. Note that the selection criteria should favor macromolecular systems where efforts are already under way but crystallization has been difficult and the results have been borderline. Some examples of classes of systems that currently meet these criteria include membrane proteins, molecular motors, biopolymer synthetic machinery (e.g., origin of replication complexes and transcriptional pre-initiation complexes). All of these systems are elaborate and fragile, which makes them difficult to crystallize unless the conditions are just right; microgravity might improve the quality of the crystals enough to lower the resolution to a level at which key structures can be discerned.

For project selection, NASA must assemble a review panel of renowned structural biologists to ensure that projects selected resonate with the community's perception of what problems are important and difficult. An international panel might serve NASA's needs best, because European and Japanese scientists often have a greater appreciation for technical and instrumental achievement and innovation. Also, more international scientists on the selection committee means that more U.S. laboratories can apply without any appearance of conflict of interest with the panel members. NASA needs to encourage the largest possible pool of candidates, not only to ensure that the highest quality applications are selected but also because the grant publicity and application process will educate a broader research community about the capabilities of the equipment available for microgravity research.

NASA does recognize the importance of community outreach and long-term grants, as can be seen in other NASA programs. In the Life Sciences Division, a memorandum of understanding with NIH created a set of grants that could be applied for by NIH grantees, and the applications were reviewed by the NIH study sections. This program introduced NASA to a new community and educated the NIH grantees about the goals and hardware of NASA life sciences research programs. In astrophysics, NASA has given out a series of 5-year grants, valued at roughly $250,000 per year, to encourage long-term projects that either exploit NASA's technical capabilities in this field or investigate theoretical issues that will affect future equipment needs in astrophysics.

The projects funded by the grant program proposed by the task group should address many of the uncertainties that have plagued the NASA protein crystal growth program so far. By using the ISS for a long-term, reliable microgravity environment, by comparing space-grown crystals to the best Earth-grown crystals available, and by focusing on challenging systems and "hot" scientific problems, the investigators should be in a position to produce the data that the research community has been seeking about crystal growth in microgravity. The results of these cutting-edge research projects should provide a definitive answer as to whether higher quality crystals can be grown in microgravity than by using the best technologies available on Earth. If none of the projects produce a space-grown crystal that enables a breakthrough for structure determination of a biologically important macromolecular assembly, NASA should be prepared to terminate its protein crystal growth program.

However, if the projects supported by this high-profile, nationwide series of grants succeed in validating the use of crystallization in microgravity to tackle important and challenging problems in biology, demand for the facilities on the ISS can be expected to increase. At that time, NASA should develop an external user program (similar to synchrotron user programs) in which a peer-review committee selects the projects. The committee would include NASA staff on an ex-officio basis to provide information about the feasibility of proposed experiments and advice about equipment capabilities. The reviewers would evaluate applications and select participants in the program so that when a flight opportunity arises, NASA could coordinate the external users and match accepted proposals with the most appropriate available hardware. Information about the equipment for the ISS must be available to all the external users so that (1) they can participate in the decision about which hardware best fits the needs of their experiment and (2) they have confidence in the equipment and will entrust their samples to it.

Recommendation: NASA should fund a series of high-profile grants to support research that uses microgravity to produce crystals of macromolecular assemblies with important implications for cutting-edge biology problems. The success or failure of these research efforts would definitively resolve the issue of whether the microgravity environment can be a valuable tool for researchers and would determine the future of the NASA protein crystal growth program.

Throughout this report, the task group has emphasized NASA's need to demonstrate that the microgravity environment can be a valuable tool for determining macromolecular structures and will therefore have a major impact on key biological questions. This focus is a reflection of the unique attitude of the scientific community engaged in protein crystal research. In cell science, as well as in other areas of microgravity research, such as materials science, the NASA program is geared toward supporting basic curiosity-driven research, and the research communities embrace that approach. Among investigators studying macromolecular assemblies, however, the main goal is solving structures of interesting crystals, and these scientists believe that the value of NASA's program should be measured in terms of the quality of service provided to the research community; in short, does microgravity directly affect their efforts to determine structures?

While continued work on understanding the crystal growth process could bear dividends, it is of secondary importance to scientists at this time. However, if a crystal grown in microgravity turns out to be better than the best version of that crystal produced on Earth, it is certainly worth asking the researcher to investigate exactly which characteristics of the microgravity environment produced the positive result so that attempts could be made to reproduce those characteristics in ground-based laboratories. NASA could make extensions of funding for projects contingent on the researchers trying to tackle these questions, as understanding the mechanisms at work would increase the likelihood of future success. The sort of issues that would arise in such research, such as the impact of microgravity on convection and sedimentation, would provide an excellent opportunity for collaboration and coordination between scientists funded by NASA's biotechnology section and investigators supported by other parts of the Microgravity Research Division, such as the materials science program. Mechanisms to encourage such cooperative projects should be considered, as well as ways to increase communication between NASA staff whose programs may be tackling fluid behavior in microgravity from a variety of directions. Possible approaches include newsletters devoted to general findings or the formation of a committee to discuss which results have the potential to cross disciplinary boundaries.

CELL SCIENCE

Cooperation with NASA's Life Sciences Division and with Other Federal Agencies

As NASA looks to involve a broader community in its cell science programs, it would make sense to take advantage of other federal agencies' existing relationships with cell scientists in a variety of fields. As mentioned above, joint grant solicitations with the NSF, NIH, or the Department of Energy might be effective. In addition, the work in cell science could be more closely coordinated with NASA's own Life Sciences Division. It is recommended that joint NRAs be established or incentives offered to encourage applications linking in vitro and in vivo work through cross-program collaborations. The cell science work could also benefit from interactions with the fluid physics program in NASA's Microgravity Research Division. Expertise in fluid mechanics might enable better understanding and control of various mechanical factors, such as shear stress and convective flow, that affect the culture environments.

NASA has already built a very productive relationship with NIH based on the development and use of rotating-wall vessels. The NASA/NIH Center for Three-Dimensional Tissue Culture was started in 1994 to expose a wider community to bioreactor technology by allowing researchers from government agencies (e.g., NIH, the Food and Drug Administration, and the Department of the Navy) to test new model systems for biomedical research and basic cell and molecular biology in the rotating-wall vessel hardware with technical assistance from experienced NASA personnel (NASA, 1995). Phase I projects allow researchers to use the bioreactors at the NASA/NIH tissue culture center to see if the technology might be useful; Phase II projects involve grants that allow the researchers to purchase their own bioreactors and refine their research in this equipment. This approach has been an effective way to allow the community to familiarize itself with the possibilities of the rotating-wall vessel bioreactors. The task group believes that this outreach program is an excellent idea and recommends that a wider range of investigators be reached by expanding the Phase I part of this program to include extramural (non-government) researchers.

Continuing ground-based research will play a critical role in helping to define the key biological questions most likely to be answered via space-based experimentation. The first steps include determining methodologies that can distinguish between the effects of microgravity on cells and on the cell culture environments and adapting innovative analytical equipment and culture instrumentation for space use. When enough information and appropriate technologies have been gathered to define specific areas in cell biology in which NASA's "big breakthrough" might occur, the NASA cell science program might consider launching a special program of significant grants similar to the one described in the section "Funding Research on Biologically Challenging Problems" from the protein crystal growth section of this chapter. The goals of this program would be in-depth study of particular mechanisms and careful comparison between ground-based and space-based experimentation to determine how the microgravity environment offers opportunities for investigation and knowledge that are unobtainable on Earth.

Resource Management and Communication in Times of Crisis

The need for a strong scientific voice in operational decisions must be stressed. Someone trained in biology and engineering should be made responsible either for maintaining the cultures on the ISS or for interacting, from the ground, with both the investigator and the responsible astronaut. Placing someone with a scientific background and perspective in a key decision-making position will enable more effective communications when experimental results dictate changes in the sampling or reagent-manipulation protocols. Additionally, strengthened communications will be very important when reacting to shortages in crew time or ISS resources (such as power). It is crucial that there be a coordinated effort between investigators and operations personnel to use resources effectively, giving each experiment the highest probability of success (NRC, 1998). In this regard, it may help to establish a liaison position staffed by a bioengineer who can bridge the communications gap. Currently, NASA is considering creating the position of lead increment scientist to coordinate scientific input into engineering decisions and decisions on the distribution of scarce resources (power, crew time, etc.). Serious problems can arise during a crisis if there is a nonscientist in the decision-making process. While crew safety and ISS operational integrity obviously have to come first, it is important that experimental payloads not be shut down when there are other options, just because shutting down is the easiest thing to do. NASA has historically demonstrated a certain rigidity about protocols for such things as the timing of missions, the manifesting of various experiments, and access to astronauts, and this rigidity has had a negative impact on investigators' ability to carry out experiments. In the future, decision-making processes should be more flexible to accommodate the needs of biotechnology researchers. The research community would be reassured by seeing NASA place bioengineers and biological scientists with the appropriate appreciation of research goals and scientifically oriented reflex responses in high enough decision-making positions to ensure that research opportunities are optimally utilized.

Bibliography

Ansari, R., S. Kwang, A. Arabashi, W. Wilson, T. Bray, and L. DeLucas. 1997. A Fiber Optic Probe for Monitoring Protein Aggregation, Nucleation and Crystallization, *J. Cryst. Growth* 168: 216-235.

Arndt, U.W. 1990. Focusing optics for laboratory sources in X-ray crystallography, *J. Appl. Cryst.* 23: 161-168.

Arndt, U.W., P. Duncumb, J.V.P. Long, L. Pina, and A. Inneman. 1998a. Focusing Mirrors for Use with Microfocus X-ray Tubes, *J. Appl. Cryst.* 31: 733-741.

Arndt, U.W., J.V.P. Long, and P. Duncumb. 1998b. A Microfocus X-ray Tube Used with Focusing Collimators, *J. Appl. Cryst.* 31: 936-944.

Brown, D.D., U.W. Goodenough, S.C. Harrison, A.P. Mahowald, E.M. Meyerowitz, C.R. Somerville, and A.L. Staehelin. 1998. Report on NASA Life Sciences Research to the American Society of Cell Biology Council. Available at <http://www.ascb.org/ascb/pubpol/nasareport.html>.

Carter, D.C., B. Wright, T. Miller, J. Chapman, P. Twigg, K. Keeling, K. Moody, M. White, J. Click, J.R. Ruble, J.X. Ho, L. Adcock-Downey, G. Bunick, and J. Harp. 1999a. Diffusion-controlled crystallization apparatus for microgravity (DCAM): flight and ground-based applications, *J. Cryst. Growth* 196: 602-609.

Carter, D.C., B. Wright, T. Miller, J. Chapman, P. Twigg, K. Keeling, K. Moody, M. White, J. Click, J.R. Ruble, J.X. Ho, L. Adcock-Downey, T. Dowling, C.H. Chang, P. Ala, J. Rose, B.C. Wang, J.P. Declercq, C. Evrard, J. Rosenberg, J.P. Wery, D. Clawson, M. Wardell, W. Stallings, and A. Stevens. 1999b. PCAM: A multi-user facility-based protein crystallization apparatus for microgravity, *J. Cryst. Growth* 196: 610-622.

Chayen, Naomi E., and John R. Helliwell. 1999. Space-grown crystals may prove their worth, *Nature* 398: 20.

Cogoli, A., and M. Cogoli-Greuter. 1997. Activation of lymphocytes and other mammalian cells in microgravity, *Adv. Space Biol. Med.* 6: 33-79.

Day, J., and A. McPherson. 1992. Macromolecular crystal growth experiments on International Microgravity Laboratory, *Protein Sci.* 1: 1254-1268.

DeLucas, L., C.D. Smith, H.W. Smith, S. Vijay-Kumar, S.E. Senadhi, S.E. Ealick, D. Carter, and A. McPherson. 1989. Protein crystal growth in microgravity, *Science* 246: 651-654.

Dickson, K.J. 1991. Summary of biological spaceflight experiments with cells, *ASGSB Bull.* 4: 151-260.

Dobrianov, I., C. Caylor, S.G. Lemay, K.D. Finkelstein, and R.E. Thorne. 1999. X-ray diffraction studies of protein crystal disorder, *J. Cryst. Growth* 196: 511.

Dobrianov, I., K.D. Finkelstein, S.G. Lemay, and R.E. Thorne. 1998. X-ray topographic studies of protein crystal perfection and growth, *Acta Crystallogr.* D54: 922.

Freed, L.E., Robert Langer, Ivan Martin, Neal R. Pellis, and Gordana Vunjak-Novakovic. 1997. Tissue engineering of cartilage in space, *Proc. Natl. Acad. Sci. USA* 94: 13885-13890.

Geierstanger, B.H., M. Mrksich, P.B. Dervan, and D.E. Wemmer. 1996. Extending the recognition site of designed minor groove binding molecules, *Nature Structural Biology* 3: 321-324.

Gilliland, G.L., M. Tung, D.M. Blakeslee, and J.E. Ladner. 1994. The biological macromolecule crystallization database, version 3.0: new features, data, and the NASA archive for protein crystal growth data, *Acta Crystallogr.* D50: 408-413. See also <http://www.rcsb.org/pdb/>.

Hammond, T.G., F.C. Lewis, T.J. Goodwin, R.M. Linnehan, D.A. Wolf, K.P. Hire, W.C. Campbell, E. Benes, K.C. O'Reilly, R.K. Globus, and J.H. Kaysen. 1999. Gene expression in space, *Nature Medicine* 5: 359.

Henry, Celia M. 1999. The incredible shrinking mass spectrometers, *Anal. Chem.* 71: 264A-268A.

Kao, S., H. McDonald, and W. Wilson. 1998. Usefulness of virial coefficients in protein crystal growth, presented at the 7th International Conference on the Crystallization of Biological Macromolecules, Granada, Spain.

Kaysen, J.H., W.C. Campbell, R.R. Majewski, F.O. Goda, G.L. Navar, F.C. Lewis, T.J. Goodwin, and T.G. Hammond. 1999. Select de novo gene and protein expression during renal epithelial cell culture in rotating wall vessels is shear stress dependent, *J. Membr. Biol.* 168: 77-89.

Köhler, S., C.F. Delwiche, P.W. Denny, L.G. Tilney, P. Webster, R.J.M. Wilson, J.D. Palmer, and D.S. Roos. 1997. A plastid of probable green algal origin in apicomplexan parasites, *Science* 275: 1485-1488.

Koszelak, S., J. Day, C. Leja, R. Cudney, and A. McPherson. 1995. Protein and virus crystal growth on international microgravity laboratory, *Biophys. J.* 69: 13-19.

Lewis, Marian L., and Millie Hughes-Fulford. 1997. Cellular responses to spaceflight, pp. 21-39 in *Fundamentals of Space Life Sciences*, Vol. 1, Susanne E. Churchill (ed.). Malabar, Florida: Krieger Publishing Company.

Malakoff, David. 1999. A $100 billion orbiting lab takes shape. What will it do? *Science* 284: 1102-1108.

McPherson, A., A.J. Malkin, Y.G. Kuznetsov, S. Koszelak, M. Wells, G. Jenkins, J. Howard, and G. Lawson. 1999. The effects of microgravity on protein crystallization: Evidence for concentration gradients around growing crystals, *J. Cryst. Growth* 196: 572-586.

Moore, D., and A. Cogoli. 1996. Gravitational and space biology at the cellular level, pp. 1-106 in *Biological and Medical Research in Space*, D. Moore, P. Bie, and H. Oser (eds.). New York: Springer.

National Aeronautics and Space Administration (NASA). 1995. NASA, NIH sign agreement on biomedical research, *Space Technology Innovation* 3:5.

National Aeronautics and Space Administration (NASA), Office of Life and Microgravity Sciences and Applications, Human Exploration and Development of Space (HEDS) Enterprise. 1997. Microgravity Biotechnology: Research and Flight Experiment Opportunities, NRA-97-HEDS-02. Washington, D.C.: NASA.

National Research Council (NRC). 1995. Microgravity Research Opportunities for the 1990s. Washington, D.C.: National Academy Press.

National Research Council (NRC). 1997. Future Materials Science Research on the International Space Station. Washington, D.C.: National Academy Press.

National Research Council (NRC). 1998. A Strategy for Research in Space Biology and Medicine in the New Century. Washington, D.C.: National Academy Press.

National Research Council (NRC). 1999. Institutional Arrangements for Space Station Research. Washington, D.C.: National Academy Press.

Reichhardt, T. 1998. Biologists recommend scrapping NASA's research on crystals, *Nature* 394: 213.

Searby, Nancy D., Gordana Vunjak-Novakovic, and Javier de Luis. 1998. Design and development of a space station cell culture unit, presented at the Session on Environmental Considerations in Microgravity Flight Implementation II at the International Conference on Environmental Systems, Danvers, Mass.

Smith, G.D., E. Ciszak, and W. Pangborn. 1996. A novel complex of a phenolic derivative with insulin: Structural features related to the T->R transition, *Protein Sci.* 5: 1502-1511.

Snell, E.H., S. Weisgerber, J.R. Helliwell, E. Weckert, K. Holzer, and K. Schroer. 1995. Improvements in lysozyme protein crystal perfection through microgravity growth, *Acta Crystallogr.* D51: 1099-1102.

Unsworth, Brian R., and Peter I. Lelkes. 1998. Growing tissues in microgravity, *Nature Medicine* 4: 901-907.

Volkman, B.F., M.J. Nohaile, N.K. Amy, S. Kustu, and D.E. Wemmer. 1995. Three-dimensional solution structure of the N-terminal receiver domain of NtrC, *Biochemistry* 34: 1413-1424.

Appendixes

A

Hardware Available or in Development and Schedule for Biotechnology Research on the International Space Station

Research equipment on the ISS will be housed in racks, as it was on the space shuttle. For the ISS, there are two kinds of racks: EXPRESS (Expedite the Processing of Experiments to Space Station) racks and ISPR (International Standard Payload Rack). ISPRs are the basic housing within the various modules (U.S., Japanese, Russian, etc.) that make up the ISS. ISPRs can then be fitted with specialized racks designed to support specific projects or research areas; the planned Biotechnology Facility (BTF) is one such rack. Alternatively, the EXPRESS racks, which are generic experiment support structures, can be fitted into the ISPR. The EXPRESS rack has subsystems for providing experiments with necessary resources such as power, water, vacuum, and gases. The standard EXPRESS rack holds eight middeck locker equivalents (MLEs) and two drawers (for storage). MLEs are the standard unit for hardware volume on the space shuttle and the ISS, and even within customized racks, such as the BTF, experimental hardware will still consist of modular MLE units. Each MLE is 20 by 16 by 11 in., and the modular equipment that fits in the MLE can weigh approximately 60 to 70 lb. The racks are located within the cylindrical modules that make up the ISS; the hardware described below under development by NASA is all currently scheduled to be placed in the Japanese Experiment Module.

HARDWARE FOR PROTEIN CRYSTAL GROWTH IN SPACE

Basic Apparatus to House Protein Crystal Growth Hardware

Thermal Enclosure System (TES) and Single Locker Thermal Enclosure System (STES). The TES (which takes up two MLEs) and the STES (which takes up one MLE) are refrigeration and incubation modules whose internal volume temperature can be controlled to any set temperature between 4°C and 40°C. The stability of the set temperature is ±0.5°C for the STES and ±0.2°C for the TES. The thermal control is accomplished by the conduction of heat in or out of the internal enclosure through a side wall. Science hardware for biotechnology investigations can be flown inside the TES or the STES, and both systems were used to transport and house space crystallization devices during previous spaceflights. Science hardware currently flown within the TES/STES includes the DCAM, PCAM, VDAs, OPCGA, and DCPCG (see below for descriptions), as well as any new experiments requiring thermal control.

Biotechnology Ambient Generic (BAG). The BAG terminology is used to describe any flight of DCAM, PCAM, or VDA-2 hardware as stowage items subject to ambient Orbiter or ISS conditions rather than the thermally controlled environment of a TES or STES. The functionality of the DCAM, PCAM, and VDA-2 hardware is identical to its functionality when flown in an enclosure. However, the number of DCAM trays, PCAM cylinders, or VDA-2 trays flown may vary owing to differences in the ambient stowage volume. A temperature data logger is flown in conjunction with any BAG payload.

Protein Crystal Growth Hardware

Protein Crystallization Apparatus for Microgravity (PCAM). The PCAM is a cylinder with a series of nine Lexan trays positioned between interleaving actuator plates. Turning the actuator knob several revolutions to a fixed stop activates or deactivates all of the Lexan trays simultaneously. Each of the trays contains seven vapor-equilibrium chambers. In the center of each chamber is a pedestal with a depression on top that can contain up to 40 µl of premixed protein sample solution and precipitate solution. The pedestal is surrounded by a reservoir of absorbent material. The protein solution is isolated from the reservoir prior to activation and after deactivation by an elastomer. Each tray is similar to a ChrysChem apparatus, and PCAM operation is based on the principle of vapor diffusion. The PCAM allows rapid refurbishment of the hardware and experiments in the event of launch scrub turnarounds. Operations require a minimum of crew involvement and skill, and it can be operated as handheld or adapted for automation. The PCAM's disposable interface allows individual investigators to take crystals undisturbed to their respective labs for postflight analysis. The EcoRI-DNA crystals were obtained using this device. Each PCAM cylinder contains 63 individual experiments, and up to six PCAM cylinders are flown in an STES for a total of 378 chambers per MLE. PCAM was developed at New Century Pharmaceuticals (Carter et al., 1999b). Currently, work is under way on an upgrade for PCAM. Planned improvements include an increased number of samples per volume (to approximately 960 per MLE) and automated activation and deactivation of the crystallization phase. The upgraded hardware is in the early definition phase.

Diffusion-Controlled Protein Crystallization Apparatus for Microgravity (DCAM). DCAM is composed of a central housing with two reservoir chambers separated by an exchangeable gel plug of varying proportions. Each chamber is sealed by an end cap. One chamber includes a standard proportional microdialysis button, which contains the protein solution. The other reservoir chamber houses the precipitating agent, which, over time, diffuses through the gel plug. The DCAM is essentially activated when it is loaded on the ground, so no crew resources are required for activation or deactivation. DCAM operation is based on the principle of liquid-liquid diffusion. DCAM was designed for long-duration protein crystallization on the Mir space station, and the equilibration profiles are extremely stable and reliable over several months. Nucleosome core particle crystals, grown to 4 mm, were crystallized in DCAM during a 4-month mission on Mir. DCAMs are flown in tray assemblies with 27 DCAMs per tray and up to three trays per STES. This results in a total of 81 samples per MLE. DCAM was developed at New Century Pharmaceuticals (Carter et al., 1999a). Currently, work is under way on an upgrade for DCAM. Planned improvements include a larger number of samples per volume (to approximately 200 per MLE) and more crystallization options: vapor diffusion, bulk, and dialysis. The upgraded hardware is in the early definition phase.

Enhanced Gaseous Nitrogen Dewar (EGN). Because the EGN is an ambient stowage experiment, it is not housed in the thermally controlled environment of an STES. The hardware consists of a liquid nitrogen dewar and aluminum insert tube, sealed Tygon capillary sample tubes, sample bundles, and an electronic temperature monitoring system. The dewar vessel consists of two flasks with the inner space evacuated to create a thermal vacuum insulation. The inner flask contains a calcium silicate absorbent, at the center of which is a cylindrical container. The dewar insert, sample bundles, and protein sample batches are placed inside this cylinder. Liquid nitrogen is poured into the inner flask and is absorbed by the calcium silicate. Approximately 7 days after the EGN and its samples have been assembled, all the liquid nitrogen boils off, the sample proteins thaw, and protein crystal growth commences and continues inside the individual Tygon tubes for the duration of the mission. No activation or deactivation by the crew is required, and the samples are returned to the investigator immediately after landing. The EGN is the size of an MLE, and the insert that contains the sample material is 11.4 inches long by 2.95 inches in diameter. The following numbers of samples per MLE are possible, assuming that all samples are of the same size:

Sample Size (µl)	Number of Samples	Source of Estimate
10-20	10,000	Calculation
170	1,000	Calculation
600	150-180	Actual flight data

The EGN is based on previous flight experience with a similar piece of hardware known as the gaseous nitrogen dewar. Both were developed at the University of California at Irvine, and the EGN is in fabrication for flight in 2000.

Second-Generation Vapor Diffusion Apparatus (VDA-2). The VDA-2 is based on a triple-barreled syringe system designed to mix protein solutions and precipitants in microgravity. Each barrel of syringe can be loaded with up to 30 µl of solution. The experiment is initiated by deployment of the solutions onto the tips of the syringe assemblies to form drops and deactivated by moving the drops back into the syringes. Absorbent reservoir material containing approximately 1 ml of precipitant solution surrounds the drop in each chamber. Mixing of the drops is achieved by moving the droplet solutions into and out of the third syringe barrel. The VDA-2 is activated on orbit by a crew member, who operates a mechanism that injects and mixes solutions in all growth chambers simultaneously; the crew member then deactivates the entire VDA-2 before leaving orbit. Upon deactivation, the drop containing the crystal is drawn back into the syringe and the end of the syringe is plugged for subsequent recovery and delivery to the investigator. The advantage of using VDA-2 is that the solutions are mixed in microgravity and the starting point of the drop need not be in the soluble range. Up to four VDA-2 trays may be flown in an STES, for a total of 80 experiments per MLE. A commercial version of this system includes more samples (128) but does not allow for photography in orbit. VDA-2 was developed at the University of Alabama at Birmingham (DeLucas et al., 1989).

Protein Crystallization Facility. This equipment is used for batch processing of proteins whose solubility depends on temperature. Sample bottles in this device range from 50 to 500 ml, and four sample bottles can be accommodated in one STES. A temperature gradient along the linear axis of the cylinder can be manipulated by controlling the temperature of one end. Another version of this device includes laser light scattering to detect nucleation so that the temperature can be controlled manually. This version can contain two sample bottles. The hardware was developed at the University of Alabama at Birmingham.

Devices in Fabrication for in Situ Observation of Crystallization on Orbit

Interferometer for Protein Crystal Growth. This system employs a Michleson-Morley phase-shift interferometer to produce images showing density changes in solution as a protein crystal forms. The system comprises three major systems—an interferometer, six fluid assemblies with test cells, and a flight data system. The crystal growth cells are made of optical-grade glass: cells are 1 mm thick and contain 250 µl of solution. Each fluid handling system is a self-contained plastic assembly enclosing two pairs of 4-ml supply syringes (one containing protein solution and one containing precipitant solution), a waste receptacle, and a test cell, as well as mechanisms to inject fluids and to position the test cell. The crew operates the fluid system with a hand crank that depresses the syringe pistons. Six of these assemblies now exist, but any number can be reproduced. The flight data system includes a 486-based laptop computer and has video recording capability. It was originally designed to perform an experiment in the Mir glovebox and was developed at the University of California at Irvine.

Observable Protein Crystal Growth Apparatus (OPCGA). This equipment is designed to observe the formation of nutrient concentration depletion zones in the vicinity of growing crystals using a fused optics, phase shift Mach-Zehnder interferometer. The crystals are formed in 96 growth cells, each of which represents an individual liquid-

liquid diffusion crystallization experiment. The cells are mounted on rails in sets of four surrounding a central shaft carrying opposing optical systems, each of which consists of a phase shift interferometer with a field of view 4 mm by 4 mm. The optical systems will resolve an index of refraction of 1.7×10^{-5} through a 1 mm optical path. The resolution will be 15 μ. A black and white camera to make a video of the growing crystal is also available. The optical system will also contain a polarization microscope with a 4 mm by 4 mm field of view and a 1:1 magnification. The polarization microscope uses a diode for backlighting the growth cell. A color camera is included for time-lapse video. A VCR will record the video microscopy data. It is hoped that this device will produce data that can be used to compare the properties and kinetics of formation of concentration gradients in mother liquids for protein crystal growth in microgravity and in conventional laboratory environments. The OPCGA contains 96 growth cells flown within a TES, resulting in 48 samples per MLE. The OPCGA was developed at the University of California at Irvine (McPherson et al., 1999) and is in fabrication for flight in 2002.

Dynamically Controlled Protein Crystal Growth (DCPCG). The DCPCG system uses controlled dehydration or temperature as function of time to grow crystals in orbit. There are three components of the DCPCG: a V-locker, a T-locker, and a C-locker. In the V-locker, a closed loop dry nitrogen gas system controls the rate of water evaporation from protein solutions held at a constant temperature of $22\pm0.5°C$. In the T-locker, thermal fluctuation induces supersaturation of sample solutions and subsequent crystallization. The T-locker provides temperature control to sample growth chambers in the range of 4°C to 50°C, with a temperature control ramp rate of 1.0°C per min. at the low and high limits of the temperature range. In both the V- and T-lockers, there are at least 30 sample chambers, 10 of which are "control" chambers that interface with a laser light scattering system to detect sample aggregation. Each sample chamber will hold 40 to 200 μl of sample solution. A video subsystem is incorporated as an integral link for near-real-time experiment evaluation. Both the V- and T-lockers interface with the C-locker to provide external communications, video capture, and data storage. A version of this system will soon be available commercially for growing crystals in laboratories on Earth. The DCPCG contains at least 60 sample chambers and occupies a total of three MLEs, resulting in at least 20 samples per MLE. The DCPCG was developed at the University of Alabama at Birmingham and is in fabrication for flight in 2000.

Devices in Early Definition Phase for in Situ Observation of Crystallization on Orbit

Microscope Laser Light Scattering Apparatus. This device is being designed to permit scientists to determine the size and relative concentration of protein molecules attaching onto a crystal's surface. To do so, the device will use multi-angle dynamic laser light scattering and fluorescence recovery after photobleaching diagnostic techniques. Although the number of cells to be examined has not yet been determined, temperature control of each cell is postulated to be 0.1°C between 0° and 60°C. Cooling of 0.6°C per hour is anticipated. This hardware is being developed at the Naval Research Laboratory and is in the early definition phase.

Glovebox-size Interferometer. This glovebox-sized instrument will be specifically designed to study improvement of crystal quality through imposed changes in the transport conditions in the solution. Six cells with individual temperature controls to 0.01°C will be used. The cells will be capable of fixed position crystal seeding, and an automatic phase-shifting interferometer (operated from the ground) will be available for in situ surface characterization for each cell. In addition, a high-resolution digital camera will image the crystal. The intent is to resolve and study single growth steps and follow the details of step bunching. This device is being developed at University of Alabama at Huntsville and is in the early definition phase.

X-ray Crystallography Facility

This facility has been designed and is in fabrication at the Center for Macromolecular Crystallography at the University of Alabama at Birmingham. There are several components; two full racks (16 MLEs) are currently reserved for this facility, which is separate from the planned BTF on the ISS.

High-Density Protein Crystal Growth (HDPCG) Unit. This equipment contains 168 individually molded six-packs, each of which has six wells in which vapor diffusion experiments can be performed. Each well of the six-pack can hold 49 µl of protein solution and 500 µl of reservoir solution. The six-packs can be removed from the HDPCG for observation in the visualization unit (see below) or sampling in the CPPI (see below). In the current configuration, the HDPCG cannot be reloaded on orbit. Future plans call for adding on-orbit reloading capabilities, as well as the option of running liquid-liquid diffusion experiments. The unit occupies one MLE and fits 1,008 samples in that volume.

Video Command and Monitoring System. This unit provides an imaging system of still pictures with 12× magnification (not actual video) for samples in the six-packs of the HDPCG. The crystallization results can be visualized either during or after a period of crystal growth, and the resulting images may be used to select appropriate candidates for mounting or freezing in the CPPI (see below). This unit occupies one MLE and will be located near the HDPCG.

Crystal Preparation Prime Item (CPPI). This unit is designed to remove crystals from the HDPCG six-packs and mount them on hair loops for cryopreservation or hair loops inside a capillary, unfrozen. The temperature inside the CPPI can be controlled and maintained between +4°C and +30°C. The robotic arm used for all manipulations was developed under a cooperative agreement between MicroDexterity Systems and NASA's Jet Propulsion Laboratory and was originally designed with eye surgery in mind. The robot, referred to as the OM3™, has six degrees of freedom within a work volume of 400 cc. It has a resolution of 10 µ and repeatability of 25 µ in terms of positioning. At any given time, the CPPI has room for two cartridges, each holding two HDPCG six-packs, 12 hair loops or capillaries, and 9 pipette tips. Using these tools, the robot can remove crystals from a well and mount a crystal in about 5 minutes. In the future, this process will be speeded up to about 1.5 minutes, and the actual exposure of liquids to the controlled environment will be about 30 seconds. After the loop mount of the crystal, the loop and crystal are presented to a cold volume and pushed rapidly through the –183°C nitrogen gas to simulate the flow of cold nitrogen gas from a cold stream. Then, the frozen crystal is transferred to an insulated thermal mass for transfer to the X-ray diffraction system or stored in the 24-position storage freezer, also located in the CPPI. Capillary mounts bypass the freezer. Crystals stored in the storage freezer may be transferred later to the MELFI (see below).

X-ray Diffraction System. The diffraction system is divided into three parts: the goniometer, the detector, and the X-ray generator. The goniometer is a three-circle type with χ fixed at 45°. The detector is currently configured to move such that crystal to detector distance is between 60 and 200 mm and the swing angle (2θ) is ±45°. The current SMART 2K CCD detector has a 135-mm diameter imaging area with a 1.9:1 fiber optic taper for reducing the image from the phosphor to the CCD. The pixel to pixel resolution is 48 µ unbinned in the 2048 mode. The highest resolution data the detector could collect at 80 mm detector distance and swung to 45° would be 1.1 Å. The Bruker rule of thumb for low-divergence X-ray beams is that the longest unit cell is equal to the detector distance divided by 0.6. This implies a 133 Å unit cell at an 80 mm detector distance and a 333 Å unit cell at a 200 mm detector distance. Plans for future improvements include using a 6K single chip CCD detector system currently under development by Bruker and redesigning the goniostat to allow the crystal to detector distance to expand to 300 mm (which would enable the system to achieve a 500 Å unit cell). The X-ray source is a Microsource system, manufactured by Bede Scientific. It is designed to be a low-power, intense X-ray source (Arndt et al., 1990, 1998a, b). The Microsource is powered at 40 kV and 0.6 mAmps or 24 W of power. It currently is about 1/3 to 1/2 as intense as the Yale/MSC type mirrors on a 5 kW Rigaku rotating anode generator. This source is stationary following fine adjustment. The X-ray diffraction system can handle crystals between 0.1 mm and about 2 mm, and the crystal can be translated ±5 mm in the x, y, and z directions and can be rotated about ϕ and Ω a full 360°.

Cryocooler System. This system employs a molecular sieve system to separate nitrogen from the air within the space station for cooling using a Stirling type cooler (capable of a 25-W heat lift at –183°C). Flow rates are currently about 3.5 to 4 liters per minute at the crystal, for both the cold and outer flows. Temperatures at the crystal position have been measured at −183°C. A new nozzle developed by Hakon Hope, which uses a heater element to produce a warm outer flow from the inner cold flow, is being tested.

Experiment Control. The status and operation of the various components of the XCF (the HDPCG, the visualization unit, the CPPI, and the X-ray diffraction system) on the ISS can be monitored on the ground through a Payload Operations Control Center. Current plans call for these centralized control capabilities to be situated at the Center for Macromolecular Crystallography at the University of Alabama at Birmingham.

Relevant Support Equipment

The apparatus listed here is not a part of the BTF or the XCF but is scheduled to be located elsewhere on the ISS. Like crew time for experimental activities, use of support equipment will be shared by many research projects on the ISS.

Minus Eighty Degree Laboratory Freezer for ISS (MELFI). This unit will provide cooling down and storage for reagents, samples, and perishable materials in four dewars with independently selectable temperatures of −80°C, −26°C, and +4°C during on-orbit ISS operations. It will also be used to transport samples to and from the ISS in a low-temperature controlled environment. The total capacity of MELFI is 300 liters; the system occupies a full rack. MELFI is currently in development by the European Space Agency; delivery is due late in 2000.

Cryofreezer System. This unit is designed to provide ultrarapid freezing and storage capacity for 3 liters of research specimens at −183°C. It is also under development by the European Space Agency for delivery in 2004.

Middeck Glovebox. This unit provides an enclosed space for experiment manipulation and observation for work in the several disciplines to be studied on ISS, including protein crystallization, fluid physics, combustion, and material science. Various modes of air circulation and pressurization are possible. Multipurpose filters are used to remove particles, liquids, and reaction gases from the circulated air. The glovebox, which occupies two MLEs, has a working volume of 35 liters and a door opening of 20.3 by 19.4 cm for sample and hardware transfer. Up to 60 W of 24, 12, and 5 V of direct current power is available for instruments to be used inside the glovebox.

HARDWARE FOR CELL SCIENCE IN SPACE

Cell and Tissue Culture Hardware in Development for ISS

Biotechnology Temperature Controller (BTC). This unit is designed to provide refrigeration on-orbit as well as the capability of preserving and incubating multiple cell cultures simultaneously. The cell culture bags are transparent to allow visualization of the samples by light microscopy. While the BTC does not have the capability of automated medium exchange, the cultures can be fed using special needleless "penetration" connectors on the bags that provide for multiple aseptic connections. The BTC can be used as one large chamber or reconfigured into 2, 3, or 4 chambers with separate environmental controls (temperatures can range from 4°C to 40°C in 1°C increments). The unit occupies one MLE and can contain up to 120 7-ml culture bags, which are Teflon sample modules containing media and cells. This unit is in development at NASA, and its first flight is scheduled for late 2001. The BTC, with its combined refrigeration and incubation capabilities, was designed based on lessons learned from NASA's Biotechnology Specimen Temperature Controller (BSTC) and Biotechnology Refrigerator (BTR), which have been flown both for short-term space shuttle trips and for long-term experiments on Mir.

Cell Culture Unit (CCU). This unit is a modular cassette-style bioreactor that can accommodate multiple cell culture chambers (see Figure 2.2). The CCU provides control of temperature (between 4°C and 39°C) and pH (between 3.5 and 8.5) and allows for continual feeding and waste medium harvest from perfused stationary cultures (Searby et al., 1998). Mixing occurs via medium recirculation. The CCU also provides automated sample collection and injection and high-quality video microscopy. Individual perfused culture chambers can be replaced on orbit. Specimens are loaded in chambers on the ground; inoculation and subculture can occur in space. Bubbles must be manually prevented from accumulating in the chambers. The CCU can accommodate from 8 large (30 ml) to 24 small (3 ml) samples and the associated support and observation equipment within 2.5 MLEs. This piece of hardware is under development by Payload Systems, Inc., in conjunction with the Massachusetts Institute of Technology for the Life Sciences Division of NASA and is scheduled for its first flight in 2002. The cell science program within the Microgravity Research Division is funding early development work on the Perfused Stationary Culture System, which is supposed to be a small-volume (5 to 50 ml), multivessel system for on-orbit cell culture and tissue engineering investigations. This system is in the early stages of development, has planned goals and capabilities similar to those of the CCU, and may not be developed if the CCU proves to be successful.

Rotating-Wall Perfused System (RWPS). This unit houses a single 125-ml rotating-wall perfused vessel in a controlled environment along with associated equipment for medium infusion/perfusion, temperature control, gas exchange, and independent rotation. Unlike ground-based rotating-wall bioreactors, in which laminar flow is set up to randomize the force vectors and to minimize the shear stress, the space-based vessels have rotating walls in order to produce Couette flow, which augments mass transport. Observation and video recording are possible through a large window in the front of the unit. The RWPS can be inoculated on the ground just before launch or on orbit, but once the RWPS is powered and the experiment initiated, it remains powered throughout the increment until landing. Cell and media samples can be removed on orbit through sample ports located on the side and front panels. The RWPS occupies one MLE and supports one cell or tissue sample. It is scheduled for its first flight late in 2000. The RWPS is an updated version of NASA's Engineering Development Unit (EDU), which has housed rotating-wall vessel experiments on the space shuttle and on Mir (see Figure 2.3).

Cell Science Support Equipment

Sensor and Control Technologies. NASA sensor research by on-site contractors focuses on fluid sensors that will enable physiological control of the cell/tissue culture media environment. Sensors for pH and glucose, as well as a pH control system, are at advanced stages of development. In contrast, sensors to measure oxygen and carbon dioxide concentrations are in the early stages of development. Sensors will be installed within cell and tissue culture hardware in order to assist in data collection and remote operation.

Gas Supply Module. This unit is designed to provide nitrogen, carbon dioxide, and other gases to experiments in the RWPS and the BTC. A completed version of this equipment has been flown on Mir, and the revised version, which is 0.5 MLEs, is scheduled for its first flight in 2002.

Experiment Control System. This unit is intended to serve as the standarized control system for biotechnology investigations. It must be able to interface with experiment-specific hardware (e.g., the RWPS) as well as general support systems (e.g. the Gas Supply Module). Tasks include data acquisition and archiving, experiment control, and communication with the ground and the rack systems (e.g., power supply). The Experiment Control System occupies half of an MLE.

Hydrodynamic Focusing Bioreactor (HFB). This bioreactor is a variation on the standard rotating-wall vessel central to the EDU and the planned RWPS. The HFB was designed to control the directional movement and removal of air bubbles from the bioreactor vessel on orbit without degrading the low-shear culture environment or the tissue assemblies. The HFB produces a low-shear fluid environment, while a variable hydrofocusing force is used to control the movement, location, and removal of suspended cells, tissue aggregates, and air bubbles from the reactor. A space version of this hardware is currently in the design and testing phase.

General Support Equipment Relevant to Cell Science Research

The apparatus listed here is not a part of the BTF but is scheduled to be located elsewhere on the ISS. Like crew time for experimental activities, use of support equipment will be shared by many research projects on the ISS.

Minus Eighty Degree Laboratory Freezer for ISS (MELFI). This unit will provide cooling down and storage for reagents, samples, and perishable materials in four dewars with independently selectable temperatures of –80°C, –26°C, and +4°C during on-orbit ISS operations. It will also be used to transport samples to and from the ISS in a low-temperature controlled environment. The total capacity of MELFI is 300 liters; the system occupies a full rack. MELFI is currently in development by the European Space Agency; delivery is due late in 2000.

Cryofreezer System. This unit is designed to provide ultrarapid freezing and storage capacity for 3 liters of research specimens at –183°C. It is also under development by the European Space Agency; delivery is due in 2004.

Water and Air Delivery Systems. These systems are in the early stages of development. A water sterilization and filtration system is planned to allow shuttle and ISS water to be purified and used for media preparation (rehydration/dilution) for cell culture. An air purification system would allow the use of shuttle and ISS cabin air to aerate cell culture medium through a separately housed oxygenator. Such air would be enriched with up to 10% carbon dioxide.

Incubator. This is a controlled environmental chamber for growing cell and tissue cultures. Its available capacity is 18.7 liters and the temperature range is 4°C to 38°C. It operates within the Life Sciences Glove Box or the Centrifuge Rotor. The glove box provides an enclosed space for experiment manipulation and observation for life sciences research on the ISS. Its volume is 500 liters, it can accommodate two habitats, and two crew members can conduct scientific procedures simultaneously. It is scheduled for launch in 2001. The Centrifuge Rotor is a 2.5-m centrifuge designed to provide a simulated gravity environment from 0.01 to 2.00 times Earth's gravity. The habitat volume available within the centrifuge is approximately 0.18 m^3. Currently, the planned launch date for the centrifuge is 2004.

Analytical Equipment

A variety of analytical equipment is scheduled to be available on ISS. Some instruments, such as the cameras, will be easily transportable throughout the ISS; others, such as the microscopes, will be operated within the glove boxes, and still others will be located within their own modules. Current plans call for the microscopes to be linked to digital cameras.

Cameras. One commercial off-the-shelf (COTS) digital still camera, two COTS film still cameras, and a COTS camcorder for the general-purpose videotaping of experiments.

Dissecting Microscope. This instrument allows for microscope-aided inspection and manipulation of specimens within the confines of a glove box. Magnification range is 4 to 120×.

Compound Microscope. This is a standard benchtop microscope with magnification up to 1000× and Kohler illumination to support phase-contrast microscopy for cellular and subcellular observations. It will use halogen, mercury, or xenon light sources.

Miscellaneous

Mini-Payload Integration Center (Mini-PIC). This system is a fairly portable ground-based hardware system that emulates the proposed configuration and capabilities of the BTF rack. It has been developed to enable principal investigators to develop protocols and carry out duplicate control experiments in their own labs.

SCHEDULE

All dates offered below are approximate and are based on the schedule provided to the Task Group by NASA in the summer of 1999.

Phase I (2000 through mid-2003). Biotechnology research on the ISS occurs using instruments that already exist and were used or planned for use on the space shuttle. These instruments will be installed in EXPRESS racks in whatever laboratory modules have been completed. These preexisting pieces of hardware were not designed for long-term flight and often do not fulfill ISS requirements for configuration and resource use, but they will receive waivers to allow continuation of the science program while new hardware is developed and constructed. The performance of hardware during this phase will provide valuable input to the design and development of ISS-specific equipment. Another constraint on scientific work done at this time will be the ongoing construction of the ISS, which will limit both the time crew can be involved in research and the available transport volume on the shuttle. At this point, ISS resources, such as power, will also be limited and may not be reliable.

Phase II (mid-2003 through 2005). The boundary between Phase I and II is not a distinct one. Instead, there will be a gradual transition in the type of biotechnology instrumentation flown on the ISS. Phase II instrumentation consists of modular units that have been designed specifically to be used for long-duration (several months or more) experiments on the ISS. These instruments can be expected to have fewer performance risks and to meet ISS hardware requirements. Also, new efficiencies and capabilities should have been added to the equipment. For example, exchanging various sample units within hardware was not vital on a 2-week shuttle mission but would be necessary to maximize the science return from a long-duration flight. Another difference between Phase I and Phase II is that although the ISS will still be under construction, the availability of launch volume and on-orbit resources should increase. Equipment would still be housed in EXPRESS racks, but at that point both the U.S.-provided laboratory module and the Japanese-provided module should be operating.

Phase III (2005 and beyond). The transition to Phase III is sharply defined by the installation of the specialized BTF on ISS. This facility will still accept modular hardware of the type used in Phase II, but the facility will provide additional support capabilities designed specifically to enable biotechnology work on ISS. Current plans call for the BTF to provide each experimental module within it with power, gases (such as nitrogen and carbon dioxide), thermal cooling, data acquisition, storage and processing, video and image analysis, data downlink, real-time control, resource allocation, research-grade water, and vacuum exhaust for one modular unit at a time. Also, BTF will have an intelligent power distribution system to allow for efficient management of this scarce resource and to ensure that cuts in power are consistent with the constraints of each payload (i.e., power will be cut first to those payloads that are able to recover from the power outage without a sustained loss in performance). As well as using all of BTF, biotechnology experiments will also continue to operate in EXPRESS racks on ISS, when volume is available.

B

Biographical Sketches of Task Group Members

PAUL B. SIGLER *(1934-2000), Chair,* was the Henry Ford II Professor of Molecular Biophysics and Biochemistry at Yale University and an investigator of the Howard Hughes Medical Institute. His research focused on macromolecular crystallography and structure, genetic regulation, transmembrane signaling, and chemical mechanisms in cellular regulation. Dr. Sigler was a member of the National Academy of Sciences and a fellow of the American Academy of Arts and Sciences.

ADELE L. BOSKEY is director of research at the Hospital for Special Surgery and professor of biochemistry and cell and molecular biology at the Cornell University Medical College. She investigates calcium phosphate crystal deposition within the extracellular matrices of bones, teeth, ligaments, and tendons in mammals using solution, cell culture, and in vivo models. Dr. Boskey has had experiments fly on the space shuttle in 1994 and 1996, and has served on NIH-NASA advisory panels. She is a past president of the Orthopaedic Research Society.

NOEL D. JONES retired from Eli Lilly and Company as a research advisor and group leader of macromolecular structure research, after 27 years with the company. He has many years of experience in macromolecular crystallography research, drug design, and research management. His special expertise is in the development of instrumentation for growing protein crystals and for rapid diffraction data collection. He later served as vice president of Drug Design for the Molecular Structure Corporation. Dr. Jones has frequently served on NIH, NASA, and DOE review panels for the evaluation of research programs.

JOHN KURIYAN is a professor in the Laboratories of Molecular Biophysics at the Rockefeller University and an investigator of the Howard Hughes Medical Institute. He conducts research on the structural biology of cellular signaling and DNA replication signal transduction. He received the Young Investigator Award from the Protein Society in 1997 and the Eli Lilly Award from the American Chemical Society in 1998. He is currently on the editorial boards of *Structure* and *Cell*.

WILLIAM M. MILLER is a professor of chemical engineering at Northwestern University. He specializes in cell and tissue culture applications in biotechnology and medicine. Particular areas of research include therapeutic proteins, ex vivo expansion of hematopoietic cells, and epithelial cell models of in vivo responses. He has received a Presidential Young Investigator Award from the National Science Foundation and has chaired the Food, Pharmaceutical, and Bioengineering Division of the American Institute of Chemical Engineers.

MICHAEL L. SHULER is a professor of chemical engineering at Cornell University. He is an expert in bioengineering and performs research in the areas of heterologous protein expression systems, cell culture analogs for pharmocokinetic models, in vitro toxicology, plant cell tissue culture, biodegradation, and bioremediation. Dr. Shuler is a member of the National Academy of Engineering, a fellow of the American Institute for Medical and Biological Engineering, and a member of the American Academy of Arts and Sciences.

GARY S. STEIN is professor and chairman of cell biology at the University of Massachusetts Medical School, and deputy director for research at the University of Massachusetts Cancer Center. He focuses his research on regulation of cell cycle and tissue-specific genes controlling proliferation and differentiation in normal and tumor cells.

BI-CHENG WANG is a professor of biochemistry and molecular biology at the University of Georgia. He conducts research on the structure and function of biological macromolecules, including bacteriophage and *E. coli* RNA polymerases. Dr. Wang is the past president of the Diffraction Society and co-editor of *Crystallographic Reviews*. Before coming to Georgia in 1995, he was a professor of crystallography and biological sciences at the University of Pittsburgh. He has been involved in experiments aboard the space shuttle and the Russian space station Mir that aimed to grow low-defect macromolecular crystals in microgravity.

C

Statement of Task

The purpose of this study is to evaluate NASA's plans for a Biotechnology Facility (BTF) envisioned for the International Space Station and to review the process for solicitation, selection, and experiment development to be carried out in the planned facility. Specifically, the study will:

1. Review NASA's plan for a Biotechnology Facility and evaluate the facility's potential to support the range of investigations in cell biology (cell culture, growth, and differentiation) and molecular structure (growth of biological macromolecular crystals such as proteins and nucleic acids) envisioned by NASA for near-term (i.e., up to 5 years) space station utilization. To the extent feasible, the potential of the planned BTF to support research in other areas will also be evaluated.

2. Based on current trends in science, identify design limitations that might constrain long-term (i.e., beyond 5 years) use of the planned BTF for experiments in cell biology and molecular structure. If warranted, recommend possible improvements to enhance design flexibility and expand research capability.

3. Review NASA's process for solicitation, selection, and post-selection reviews (in preparation for flight) of biotechnology flight experiments. Where warranted, recommend changes in the process that could lead to improvements in scientific quality of experiments flown in the BTF and its utilization by a broader research community.

The product of the study will be a report that evaluates NASA's plans for a Biotechnology Facility envisioned for the International Space Station and reviews the process for solicitation, selection, and experiment development to be carried out in the planned facility, and make recommendations for improvements, when warranted. Funds for this activity have been provided by NASA.

D

Glossary

Bioreactor: See rotating-wall vessel.

Biotechnology Facility: A specialized facility scheduled to be installed on the International Space Station in 2005. Current plans call for it to support both cell science and protein crystal growth research.

Biotechnology Temperature Controller: A combination refrigerator/incubator with the capability to preserve and incubate multiple cell cultures simultaneously.

Cell Culture Unit: A modular cassette-style bioreactor that can accommodate multiple cell culture chambers. It is under development at MIT for the NASA Life Sciences Division.

Crystal Preparation Prime Item: Unit containing robotic instrumentation that harvests, mounts, and freezes protein crystal samples. Part of the X-ray Crystallography Facility for ISS.

Discipline Working Group: Advisory panel for NASA on the scope of research announcements, peer-review practices, and future programmatic directions.

EXPRESS racks: Basic support structures to house experimental equipment on the ISS and to transport instruments and samples on the space shuttle. EXPRESS racks are about 80 inches tall by 33 inches deep by 41 inches wide (roughly the size of a REVCO freezer) and can provide basic resources such as power to a variety of modular experiments. Each EXPRESS rack contains eight middeck locker equivalents and two storage drawers.

Increment: The length of time between shuttle flights that transfer research equipment and samples to the ISS. For research planning purposes, this is assumed to be approximately 100 days.

Incubator: A device which maintains controlled environmental conditions, especially providing warmth for cultivation of cell cultures or coolness for preservation of cell culture samples.

Life Sciences Division: The unit at NASA that aims to define, direct, support, and evaluate science and technology programs in space life sciences to enable the human exploration and development of space. Main focus is on the fundamental role of gravity, cosmic radiation, isolation and confinement on the vital biological, chemical, physical and psychological processes of living systems in space, on other planetary bodies and on Earth.

Macromolecule: A macromolecule is a polymer, especially one composed of more than 100 repeated monomers (which are single chemical units). This report focuses on biological macromolecules, such as proteins, proteins with DNA, and RNA.

Medium: A culture or growth medium contains nutrients and provides a favorable environment for the growth of cells or tissues.

Microgravity: When biotechnology experiments are performed on the space shuttle or the International Space Station, the gravitational force experienced by the cells or the protein crystals is not exactly zero, but it is much less than the gravitational force experienced on Earth. On the ISS, researchers expect to be able to conduct experiments for which the gravitational force is approximately one-millionth of what it is on Earth.

Microgravity Research Division: The unit at NASA that contains research on the behavior of systems in the microgravity environment. Includes five disciplines: Biotechnology, Combustion Science, Fluid Physics and Transport Phenomena, Fundamental Physics, and Materials Science. The Biotechnology program contains the work on protein crystal growth and cell science discussed in this report.

Middeck Locker Equivalent: Standard size of transport and experimental units on the space shuttle and the space station. Each MLE is 18.2" wide by 10.7" high by 20.4" deep, and the modular equipment that fits in the MLE can weigh approximately 60-70 pounds. Eight MLEs fit in each EXPRESS rack.

Mosaicity: A measure of the misalignment between small coherent blocks of individual molecules within a protein crystal. Lower mosaicity results in higher quality X-ray diffraction data.

NASA Research Announcement: Solicitation and instructions for grant applications for NASA funding for ground and flight experiments. Released approximately every two years.

Resolution: A measure of how much detail can be obtained from X-ray diffraction data used for structure determination of protein crystals. The resolution, typically measured in angstroms, is specifically the minimal Bragg spacing to which diffraction measurements can be obtained. The smaller the resolution is, the more details are revealed about the protein structure.

Rotating-Wall Perfused System: Instrument that houses a rotating-wall vessel and associated support equipment for experiments on the ISS.

Rotating-Wall Vessel: Designed for culturing cells in a low-shear, low-turbulence environment. On Earth, the rotating walls set up laminar flow in order to randomize the force vector felt by the cell or tissue cultures. In space, the rotation produces Couette flow to augment mass trasport. Also known as a bioreactor.

Space Products Development Office: Responsible for NASA's commercial product development program to encourage and increase the United States' industry involvement and investment in space-based materials processing and biotechnology technologies. Oversees the Commercial Space Centers, such as the Center for Macromolecular Crystallography at the University of Alabama at Birmingham. Responsible for the XCF, which is in development at this center.

Synchrotron: A facility that produces high energy X rays by accelerating electrons along a circular path. These X rays are used for diffraction studies of macromolecular crystals to gather data that can be employed to determine the macromolecule's structure.

X-ray Crystallography Facility: A multipurpose facility designed to provide and coordinate all aspects of protein crystal growth experiments on ISS: sample growth, monitoring, mounting, freezing, and X-ray diffraction. In development at the Center for Macromolecular Crystallography at the University of Alabama at Birmingham.

E

Acronyms and Abbreviations

ARIS	Active Rack Isolation System
BAG	Biotechnology Ambient Generic
BTC	Biotechnology Temperature Controller
BTF	Biotechnology Facility
CCDPI	Command, Control, Data Prime Item
CCU	Cell Culture Unit
COTS	commercial off-the-shelf
CPPI	Crystal Preparation Prime Item (system)
DCAM	Diffusion-Controlled Protein Crystallization Apparatus for Microgravity
DCPCG	Dynamically Controlled Protein Crystal Growth
DOE	Department of Energy
DWG	Discipline Working Group
EDU	Engineering Development Unit
EGN	Enhanced Gaseous Nitrogen dewar
ESA	European Space Agency
EXPRESS	EXpedite the PRocessing of Experiments to Space Station (EXPRESS) rack
HDPCG	High Density Protein Crystal Growth unit
HFB	Hydrodynamic Focusing Bioreactor
ISPR	International Space Station Payload rack
ISS	International Space Station
IUCr	International Union of Crystallography
IWG	Investigators Working Group
MLE	middeck locker equivalent
MELFI	Minus Eighty Degree Laboratory Freezer for ISS

NASA	National Aeronautics and Space Administration
NIH	National Institutes of Health
NRA	NASA Research Announcement
NSBRI	NASA Space Biomedical Research Institute
NSF	National Science Foundation
OLMSA	Office of Life and Microgravity Sciences and Applications
OPCGA	Observable Protein Crystal Growth Apparatus
PCAM	Protein Crystallization Apparatus for Microgravity
RWPS	Rotating-Wall Perfused System
STES	Single Locker Thermal Enclosure System
STMV	Satellite Tobacco Mosaic Virus
TES	Thermal Enclosure System
VDA-2	Second Generation Vapor Diffusion Apparatus
XCF	X-Ray Crystallography Facility
XDPI	X-Ray Diffraction Prime Item